Advances in Rice Cultivation
A Complete Guide on Rice

The Editors

Dr. K.A. Ponnusamy obtained his M.Sc.(Ag.) and Ph.D. in Agricultural Extension from Indian Agricultural Research Institute, New Delhi. He was a gold medallist and best Ph.D. student of IARI. Presently, working as the Director of Extension Education, TNAU, Coimbatore. Earlier he worked as Deputy Director (Extension) at National Institute of Agricultural Extension Management (MANAGE), Hyderabad. He has more than 30 years of experience in Teaching, training, research and extension. Taught more than 50 courses in Agricultural Extension and trained more than 2000 extension workers/scientists on various subjects in Agricultural Extension in India. Published many papers in reputed National and International journals and edited few books in Agricultural Extension.

Dr. C. Karthikeyan completed his M.Sc.(Ag.) and Ph.D., in Agricultural Extension during 1992-97 at Tamil Nadu Agricultural University (TNAU), Coimbatore. Won Post Doc Fellowship to work on a project at Rockefeller Centre, Italy during 2006. Completed Australian Government's Competitive Post Doc. Fellow at Charles Sturt University, Australia during 2009-10. Graduated MBA (HRM) during 2010 at Bharathiar University, Coimbatore. He is certified by AIIMA, New Delhi as an accredited management teacher in HRM.

He Started his career as Scientist (Ag. Extension) at Central Institute of Fisheries Technology, ICAR, Cochin and later joined as Assistant Professor (Ag. Extension) at Tamil Nadu Agricultural University, Coimbatore. Currently, working as Professor (Ag. Extension) and Coordinator of KVKs in TNAU, Coimbatore.

He published 16 standard books (with ISBN) in various subjects of Agricultural Extension for the benefit of UG/PG students in Agricultural Extension, 14 book chapters, 16 technical books based on research projects, 104 scientific and popular articles in International and National journals of repute.

Advances in Rice Cultivation

A Complete Guide on Rice

— Editors —

K.A. Ponnusamy

C. Karthikeyan

2017

Daya Publishing House®

A Division of

Astral International Pvt. Ltd.

New Delhi – 110 002

Publisher's Note:

Every possible effort has been made to ensure that the information contained in this book is accurate at the time of going to press, and the publisher and author cannot accept responsibility for any errors or omissions, however caused. No responsibility for loss or damage occasioned to any person acting, or refraining from action, as a result of the material in this publication can be accepted by the editor, the publisher or the author. The Publisher is not associated with any product or vendor mentioned in the book. The contents of this work are intended to further general scientific research, understanding and discussion only. Readers should consult with a specialist where appropriate.

Every effort has been made to trace the owners of copyright material used in this book, if any. The author and the publisher will be grateful for any omission brought to their notice for acknowledgement in the future editions of the book.

Cataloging in Publication Data--DK
Courtesy: D.K. Agencies (P) Ltd. <docinfo@dkagencies.com>

Advances in rice cultivation : a complete guide on rice /
editors, K.A. Ponnusamy, C. Karthikeyan.
      pages cm
      Includes index.
      ISBN 978-93-86071-24-8 (International Edition)

      1. Rice--India.   2. Rice--Handbooks, manuals, etc.
I. Ponnusamy, K. A., editor. II. Karthikeyan, C. (Chandrasekaran),
1970- editor.

      SB191.R5A38 2017   DDC 633.180954   23

Published by   : **Daya Publishing House®**
                                     A Division of
                                     Astral International Pvt. Ltd.
                                     – ISO 9001:2015 Certified Company –
                                     4736/23, Ansari Road, Darya Ganj
                                     New Delhi-110 002
                                     Ph. 011-43549197, 23278134
                                     E-mail: info@astralint.com
                                     Website: www.astralint.com

TAMIL NADU AGRICULTURAL UNIVERSITY

Dr. K. Ramasamy, Ph.D.,
Vice - Chancellor

Coimbatore - 641 003
Tamil Nadu, India

Foreword

Rice is vital to more than half of the world's population. It is the most important food grain in the diet of Asians, especially Indians. In India, Tamil Nadu is one of the leading rice growing States wherein rice has been cultivated from time immemorial.

The scientists of Tamil Nadu Agricultural University have developed many new varieties and hybrids besides production technologies in the past ten decades resulting in substantial increase in the yields especially during and after mid sixties.

There is an urgent need at present to ensure sustainable production of rice encountering the prevailing challenges of water scarcity, reducing labour availability and declining rice areas. These challenges have to be addressed mainly with the help of advanced technologies, techniques and approaches.

Tamil Nadu Agricultural University has been a pioneering institute in rice research and in the past one decade or so, it has brought out high yielding varieties and hybrids and technologies relating to production and value addition of rice as an outcome of a large number of research projects and schemes carried out across the State.

The present book on *"Advances in Rice Cultivation: A Complete Guide on Rice"* deals with the global perspective of rice economy, genesis and improvement of rice, landraces, technologies relating to management of nutrients, water, weeds, diseases, insects, etc.

Seed production techniques, system of rice intensification, machineries for paddy cultivation, organic rice cultivation, specialty rice of Kerala State and success stories highlighting transformation of the lives of rice growers have also been well covered in this book.

The contents of the book will be highly useful to the researchers, students, extension functionaries, NGOs and the farming community in taking forward rice research and production accomplishments to new heights in the years to come.

I congratulate the Editors of this book for their commendable efforts to compile an important book of this kind.

K. Ramasamy
Vice-Chancellor
Tamil Nadu Agricultural University

Preface

Rice has fed more people over a longer period of time than any other food crop. Rice cultivation is the principal activity and source of income for millions of households around the globe. Rice could be taken to many parts of the world due to its versatility. There are abundance of research results were available all over the world and only few books were available on rice which are also only on theoretical.

The newer dimensions of basic and advanced rice cultivation techniques, newer varieties were adding up fast from Tamil Nadu Agricultural University. Hence, there is a greater need for a complete guide on rice. The aim of the book is to compile intensive as well as exhaustive rice content which integrates advanced information in rice, which is required by Extension personnel, Scientists, Students and Farmers. The book is comprises of twenty chapters with explained in high resolution visuals in an organized manner.

The book chapters are contributed by the subject experts in the respected field and took meticulous care on adding scientific content in a simple and easy readability format.

The subject matter discussed in the book are soil, seed, plant, pest, disease, nutrient, farm mechanization and post harvest aspects of rice. More emphasis were taken to add an interdisciplinary nature involving subject experts from various disciplines *viz.*, Agronomy, Soil Science, Crop Physiology, Plant Breeding, Entomology, Pathology, Seed Technology, Engineering, Economics and Extension which should help anybody to improve their knowledge in rice with others. The book could serve both to the farmers and stakeholders as a technical reference and are easy to understand the basic and advanced concepts in rice farming. In addition, it also serves the needs of extension personnel involved in transferring the recent rice technology to the farmers.

The editors expressing sincere thanks to all the chapter contributors and scientists of Tamil Nadu Agricultural University.

K.A. Ponnusamy

C. Karthikeyan

Contents

Chapter 1

Overview of World Rice Economy: Status, Challenges and Future

K. Ramasamy

Vice-Chancellor,
Tamil Nadu Agricultural University,
Coimbatore – 641 003, Tamil Nadu
e-mail: vctnau@tnau.ac.in

Introduction

Rice is the staple food of Asia and it is central to the food security of about half of the world population. Asia accounts for more than 90 per cent of world rice production and consumption. Rice production is an important source of livelihood for around 140 million rice – farming households and for millions of rural poor who work on rice farm as hired labor. It is a strategic commodity as the overall economic growth and political stability of the region depends on an adequate, affordable and stable supply of this staple crop. Despite the substantial increase in rice production, in the wake of the Green Revolution, important challenges remain in ensuring an adequate and stable supply of this commodity affordably to poor consumers. Rice is a major food staple and a mainstay for the rural population and their food security. It is mainly cultivated by small farmers in holdings of less than 1 ha. Rice is also a "wage" commodity for workers in the cash crop or non-agricultural sectors. This duality has given rise to conflicting policy objectives, with policy-makers intervening to save farmers when prices drop, or to defend consumer purchasing power when there are sudden price increases.

Rice is vital for the nutrition of much of the population in Asia, as well as in Latin America and the Caribbean and in Africa; it is central to the food security of over half the world population, not to mention to the culture of many communities. Rice is therefore considered a "strategic" commodity in many countries and is, consequently, subject to a wide range of government controls and interventions.

During the 1990s, global rice production expanded at a rate of 1.8 per cent per year - marginally above the population growth rate. By the end of the decade, it reached 400 million tonnes (Mt) in milled equivalent. Developing countries account for 95 per cent of the total, with China and India alone responsible for over half of the world output. Most of the increase in the 1990s was sustained through productivity gains rather than land expansion. In recent years the tendency for yield growth to slacken has been the cause for concern. Furthermore, competition for basic resources (in particular, land and water) from other agricultural and non-agricultural sectors, as well as the negative environmental impacts associated with rice cultivation, are expected to pose a serious challenge to the future development of the sector.

During the 1990s, global trade in rice expanded on average by 7 per cent a year to about 25 Mt. Despite such dynamic growth, the international rice market remains thin, accounting for only 5 to 6 per cent of global output. Unlike for other bulk commodities, the international rice market is segmented into a large number of varieties and qualities, which are not easily interchangeable because of strong consumer preferences. Ordinary *indica* rices are the most commercialized (some 80 per cent of international trade by the end of the 1990s) followed by aromatic (Basmati and fragrant) rices at 10 per cent, medium rices at 9 per cent and glutinous rices at 1 per cent.

Economy of Rice Markets

Rice is among the basic food products subject to a relatively high level of policy interventions that affect farm incentives and consumer prices, in both developed and developing countries. These policy interventions are applied to both domestic markets and trade. The aims of these interventions are to increase farmer income, improve consumer welfare, increase price stability and achieve high self sufficiency. Policy tools frequently used are subsidies, trade controls and price stabilization and buffer stock operation.

The first major at liberalizing trade policy, including domestic support measures, was the WTO URAA in 1995. The basic architecture of the rice policy from that Agreement largely remains today despite the progress made recently in the Bali Ministerial Meeting. The 2007/08 rice price spikes were an important event in the evolution of rice policy. It is widely held that this event halted or even reversed the process of gradual liberalization of rice markets, with increased occurrences of antarkic policy regime among importers and more frequent export restrictions by exporters.

Because of rice's importance for food security and political stability, a significant proportion of trade is conducted by state trading enterprises, some of which are also obliged to procure or distribute rice

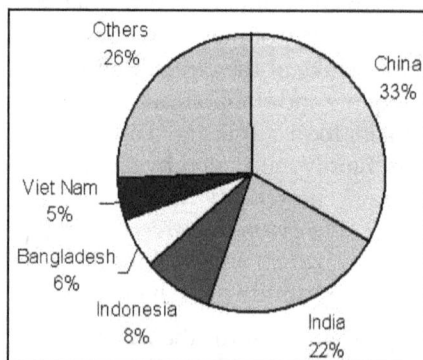

Figure 1.1: Global Rice Trade Volume and Share in Global Production.

domestically. This applies to both importing and exporting countries. However, in recent years, many of these state enterprises have lost their monopoly position and private traders have taken on greater responsibility for dealing with rice imports and exports.

Government-to-government transactions, which used to account for about half of world trade in the 1970s, are now estimated to represent less than 10 per cent of the total. In the past few years, however, they have regained popularity, as low international prices have incited or compelled governments to play a more active role in trade, either to gain bargaining power or as an indirect means of sustaining producer prices. These transactions often take place under conditions that can barely be matched by private traders, especially as far as credit is concerned.

General Thrust of Rice Government Policies

Government support to producers in developing countries concentrates mainly on: research in improved or hybrid rice varieties, investments in irrigation, preferential credits, extension and distribution of improved seed. Intervention to influence prices is also common - through procurement purchases or releases from stocks, or through changes in trade policies. In developed countries, much assistance to the sector is conveyed through direct payments and through price support. Government often plays an important role in the first phase of the marketing cycle, by procuring paddy at minimum producer prices.

In general, the involvement of the public sector in paddy processing and rice distribution is more limited. However, some governments oblige millers to purchase paddy at a predetermined price and to charge fixed mark-ups at each stage of the marketing process, while others distribute rice at fixed retail prices. It is common practice to manage rice stocks or to adopt trade policy measures in order to stabilize domestic market prices.

Trade measures, especially tariffs, are widely used to protect domestic rice markets. Despite the relatively high WTO (World Trade Organization)-bound tariffs, rice imports are often subject to "Special Safeguards" in country schedules. The role of state companies in managing international rice flows is also important, although such companies do not usually have monopoly privileges and they share their trade functions with the private sector. Many commercial transactions are conducted through government-to-government deals, the terms of which are not usually released to the market, contributing to poor transparency. Restrictions on exports of paddy or husked rice are very common, reflecting an endeavour to promote domestic rice processing. Because of the importance of rice as a staple food, many governments maintain minimum food reserves to ensure food security. In addition, countries engaged in rice distribution schemes and producer price support usually keep large rice inventories in public storage facilities. Since rice is one of the most protected traded commodities, there is considerable scope for further market liberalization. However, because of its importance in terms of food security, income generation and political stability, governments may be reluctant to loosen their control over the sector. Moreover, rice is central to the concepts of food security and multi - functionality - as promoted by a number of countries for consideration in the Doha new round of multilateral trade negotiations (launched in November 2001).

Current Issues and Problems

Falling international prices have been the principal cause for concern in the last few years, for both importing and exporting countries. The slide in world quotations was a reflection of the dynamic growth in global production since the mid-1990s, following the implementation of expansionary policies in a large number of countries. Although world paddy production has fallen in the past 2 years, supply releases from stocks have kept the downward pressure on prices. Although genetically modified rice varieties have been developed (to enhance nutritious characteristics, *e.g.* "Golden Rice", or for adaptation to extreme growing conditions, *e.g.* varieties tolerant to salty water), the issue of their acceptability worldwide has not yet gained prominence, because rice produced from such varieties are not yet widely traded. More importantly, concerns have arisen regarding the use of the "Basmati" rice denomination and claims of bio-piracy on fragrant rice genes.

Rice production sites are often the natural habitat for a wide variety of birds and plants. Water management in rice lands also ensures a soil desalination process essential to the maintenance of land fertility. As a result, environmental concerns frequently come to the fore in defense of the sector, especially in developed countries.

Domestic Support

Amber Box Domestic Support

Developed countries completed their Aggregate Measurement of Support (AMS) reduction commitments in 2000, mainly through cuts in price support. Such cuts have been associated with a rise in compensatory payments to rice producers (classified under either the "blue" or the "green" box), which have been particularly important in the EC, Japan and the United States of America. The shift from price to income aids has not been accompanied by a major fall in production and some developed countries now find themselves with large rice stocks. Support to the rice sector accounts for a very high proportion of the total AMS in Japan and the Republic of Korea. These countries are expected to resist proposals for a reclassification of policies (*e.g.* from the green or blue box to the amber box). They are also likely to oppose further reductions in the AMS on the grounds of concern for national food security and preservation of the countryside, from an environmental, cultural and social perspective.

Few developing countries have submitted a base AMS, and few are therefore subject to reduction commitments. Most developing countries still have ample scope for increasing their assistance to the sector - should they choose to do so - under the "*de minimis*" provision. Only very sizeable reductions in the *de minimis* ceilings could negatively affect rice producers in those countries where the crop accounts for an important share of total agricultural outlays, *i.e.* many Asian countries and several Latin American and Caribbean countries. The proposal to raise the *de minimis* via special and differential treatment for least developed countries may have little effect, since 10 per cent of the base production value already granted gives ample scope for domestic support to the commodity. The impact of inflation and changes in exchange rates on current AMS may be of far greater importance for countries which submitted a base AMS in domestic currencies and where inflation is high.

Blue Box Domestic Support

Decoupled, production-limiting payments are made to rice producers in the EC, Japan, the Republic of Korea and Mexico. Since 1999, they have been essential for allowing producers to weather the impact of low prices. They have been strongly criticized by other players in the rice market for not being truly "decoupled", and demands are made for their elimination or reduction by shifting decoupled income support and income safety nets from the blue or green boxes to the amber box, in order to make them subject to reduction commitments.

Green Box Domestic Support

In the United States of America, considerable resources have been channeled to the sector through production flexibility contracts, retirement payments and payments for natural disasters, all of which are classified as green box measures. Since 2002, the new Farm Bill has endogenized the counter-cyclical and emergency payments which had been previously been provided through the 1996 US Farm Bill (FAIR ACT) on an ad hoc basis; they are classified as green box measures. There is a prevailing tendency generally to promote non-commodity specific programmes, such as producer insurance schemes, also in developing countries.

A number of developing countries support the inclusion of a "Food Security" box which would permit the exemption of certain policies from reductions commitments. This would contain, *inter alia*, poverty alleviation measures and product-specific support for low-income farmers. It would be of very high relevance to rice production in a large number of countries, especially India.

Other issues

The "multifunctionality" of agriculture in terms of environmental, social and cultural concerns is being used to defend the permanence of blue and green box payments. In Japan, most of the emphasis on multifunctionality and food security is in relation to rice. In developed countries where rice is a non-marginal crop, the elimination of blue or green box support would considerably impair the sector. Rice production sites are often the natural habitat of a wide variety of birds and plants. Water management in rice lands ensures that the soil desalination process essential to the maintenance of land fertility takes place. Environmental concerns are consequently a frequently used weapon in defense of the sector.

Food safety is not particularly relevant to rice, although there is increasing concern regarding GMOs (genetically modified organisms). While some rice varieties are being developed with new genes (*e.g.* carotene-enriched rice), they are not yet traded internationally. An emerging issue which is of importance to the WTO is that of intellectual property rights over particular varieties of rice, in particular in relation to the Basmati and fragrant rice varieties developed in the

United States of America. "Biopiracy" of the genes is suspected and it is feared that the new strains could compete with traditional Basmati and fragrant rice exports from India, Pakistan and Thailand. There have also been issues regarding the use of certain denominations, such as "Basmati" or "Jasmine". India and Pakistan are now trying to have the name associated with the geographical zone of production.

International Rice Markets

Unlike maize and wheat, most rice tends to be eaten where it is produced and so does not enter international markets. Yet, the volume of international rice trade has increased almost fourfold, from 7.5 million tonnes annually in the 1960s to an average of 28.5 million tonnes during 2000–2009. In the international rice trade, a relatively small number of exporting countries must interact with a large number of importing countries. In the first decade of the 2000s, the top five exporters had 81 per cent of the world market (up from 69 per cent in the 1960s). Since the 1980s, Thailand has consistently been the world's largest exporter of rice, followed by Vietnam and India. Because of the high concentration of exports coming from only a few countries, the international rice market is vulnerable to disruptions in supply from major exporting countries, leading to higher world prices. This means that a sudden change in production trade policy in one or more of these countries could have a major impact on world market flows and prices, such as occurred in the price crisis of 2007–2008.

In contrast to exports, imports of rice are widely dispersed across countries. The five main rice-importing countries in the first decade of the 2000s (Philippines, Nigeria, Iran, Indonesia and the European Union) accounted for only 27 per cent of total global rice imports; and the share of the top 10 countries was only 44 per cent. However, because of market segmentation, some of the larger rice importers have had major impacts on world rice prices. Large purchases by state trade in the Philippines in 2007 and 2009 demonstrated how an individual importer could contribute greatly to world price destabilization. Indonesia's rice imports accounted for 10 per cent and 15 per cent of world trade in the 1960s and 1970s, respectively (and 7.4 per cent and 9.2 per cent of national net availability). During these years, Indonesia's imports had major impacts on world rice markets.

Policy

The various rice policies like price, stock and trade affect the rice sector directly as well as through their indirect effects on overall economic growth. These direct and indirect effects may benefit producers and consumers differentially and may also involve trade – offs regarding and urbanization.

Various outlooks for 2030 indicate the global demand for rice to be in the range of 503 – 544 million metric tonnes. This is equivalent to the average growth rate of approximately 1 per cent per year relative to total consumption of 439 million tonnes in 2010. This demand growth is driven mainly by the growth in population although the changing consumption pattern also has an influence. Asian rice consumption is projected to account for close to two – thirds of this total increase in demand by 2030 (GRiSP, 2010). Additional demand will arise from export markets and the projected increase in exports from Asia in 2022 relative to 2013 is in the range of 5-7 million tones, with additional imports into Africa being 2-3 million tones. Overall, the trade outlook for 2022 is of world rice trade of about 46 million tones which represents an expansion of 8-9 million tones relative to the projection for 2013.

This situation highlights the need to intensify rice production in the Asian continent to meet the rising demand as the possibility of expanding the area is limited. The projected yield growth required for meeting the increasing demand is 1.0 – 1.2 per cent per year taking into account the likely future reductions in rice area arising from competition for land from other uses. This projected growth rate is higher than the growth rate of rice yields short term vs long term impacts on food security.

Challenges Ahead

For every one billion people added to the world's population, 100 million more tons of rice need to be produced each year. But the challenges facing rice production are great. To help ensure food security, reduce poverty, and help vulnerable populations adapt to the effects of climate change, more rice needs to be produced on less land, with less water and less labor. Rice production systems need to be more equitable, efficient, environmentally-friendly, and more resilient to climate change, while contributing less to greenhouse gas emissions.

Food Security

In most of the developing world, rice availability is equated with food security and is closely connected to political stability. Changes in rice availability, and hence price, have caused social unrest in several countries. During the food crisis of 2008, the cost of rice tripled. The World Bank estimated that this pushed 100 million more people below the poverty line.

Poverty Alleviation

Of the three major crops – rice, wheat and maize – rice is by far the most important food crop for the people in low- and lower-middle-income countries. Although rich and poor people alike eat rice in low-income countries, the poorest people consume relatively little wheat and are therefore deeply affected by the cost and availability of rice. In many Asian countries, rice is the fundamental and generally irreplaceable staple, especially of the poor. For the extreme poor in Asia, who live on less than $1.25 a day, rice accounts for nearly half of their food expenditures and a fifth of total household expenditures, on an average. This group alone annually spends the equivalent of $62 billion (purchasing power parity) on rice. Rice is critical to food security for many of the world's poor people.

Shrinking Resources

Worldwide, there are more than 1508 million hectares of rice fields. Further area expansion is unlikely, but rice production must increase to meet the growing demand from a growing population. How is this possible with reduced availability of land, nutrients, water and labor, and the need to reduce the environmental footprint of rice?

Climate Change

Overwhelming scientific research and evidence have shown that the climate is changing. The vast majority of climate change impacts and the overall impact of climate change on rice production are likely to be negative. While there is still ongoing scientific exploration into climate change, IRRI recognizes two universal trends predicted by all climate change models:

☆ Temperatures will increase, resulting in more heat stress and rising sea levels, and

☆ There will be more frequent and severe climate extremes.

Tremendous opportunities exist for improving food quality, diversity and safety as well as reducing post – harvest losses through enhancements of rice value chains. Rice value chains may be either traditional or modern export chains. Traditional value chains involve rice production locally and consumed within the local production areas whereas modern export chains centres on the large urban areas which is now rapidly emerging in Asia.

Post-harvest Losses

Major opportunities for improving the value chain and reducing post harvest losses are the mechanization of post harvest operations; improving drying, storage and milling of harvested paddy; secondary processing of rice to enhance consumer convenience; the use of rice biomass to generate energy or to produce animal feed and improved vertical coordination and shortening of the rice value chain for greater efficiency. Table waste of rice (cooked rice that is never consumed) is also an increasing concern in the region. There is therefore a need to make rice available in more convenient forms through value addition to provide diversified options to rice consumers and to raise consumer awareness of table waste. There is considerable scope to improve post harvest operations and reduce the loss in harvested grain in terms of both quantity and quality.

Indian Scenario

Indian rice production has nearly trebled between 1960 and 2010, with a compounded growth rate of 2.53 per cent. Rice production in India has increased during last 60 years by about 3.5 times from 250.3 lakh tons during the first 5-yr plan period to 857.3 lakh tons during the tenth plan period. The average productivity of rice in India, at present, is 2.2 tons/ha, which is far below the global average of 2.7 tons/ha. The productivity of rice is higher than that of Thailand and Pakistan but much lesser than that of Japan, China, Vietnam and Indonesia. India is expected to surpass its demand by the year 2030, if the rice production grows at 1.34 per cent per annum. But it will remain in deficit of around 2.5 million tons, if the present growth rate of 1.14 per cent continues up to the year 2030.

The Green Revolution has helped the country to regional food surpluses with Punjab leading the country in rice production and productivity. However, despite the past achievements, rice productivity growth has to continue for obvious reasons. Looking to the future, Indian rice production will come under additional pressure from intense competition for land and water, a more difficult growing environment because of climate change, higher price for energy and fertilizers and greater demand for reduced environmental footprint. This requires a careful analysis of the current production scenario and perspectives with a view to identify researchable issues and strategies to address them.

Global demand for food is rising because of population growth, increasing affluence and changing dietary habits. The UN/FAO forecasts that global food production will need to increase by over 40 per cent by 2030 and 70 per cent by 2050 (FAO, 2009). Yet globally, water is anticipated to become scarce and there is increasing competition for land, putting added pressure on agricultural production. In addition, climate change will reduce the reliability of food supply through altered weather patterns and increased pressure from pests and diseases. Rice along with wheat forms the bedrock of Indian food security and to meet the country's stated goal of ensuring food for all, farmers will have to produce more rice from lesser land, using less water, energy and other inputs and keeping in harmony with the fragile environment. In view of the shortage of labour force at peak times and its increased cost, rice farming needs to be mechanized for which better machines are yet to be developed to reduce drudgery in rice farming and to ensure timely planting or sowing. Rice cultivation is not considered profitable and it serves more as a culture rather than agriculture as it is cultivated for domestic consumption and growing no other crop is possible in these lands during the kharif season. There is low entrepreneurship of the local population and thus low economic status.

In Tamil Nadu, 90 per cent of the farmers belong to small and marginal category and their operational holdings account 56 per cent of the total areas. So the small and marginal farmers play a key role in overall development in Agriculture and the adoption of scientific technologies by these farmers needs focused attention. The Gross Cropped Area in Tamil Nadu is around 58.43 lakh hectares of which the Gross Irrigated Area is 33.09 lakh hectares which is 57 per cent and the balance 43 per cent of the area are under rainfed cultivation. Major efforts are required to increase the productivity of rainfed crops by overcoming the various challenges such as; erratic monsoon rains, soil with low nutrient and organic contents/poor water holding capacity, soil and water erosion etc. The labour scarcity especially during the peak cropping season is also causing difficulty to the farmers to take-up timely field operations. In respect of agricultural crops, the crop cultivation is taken up in two to three seasons annually. Hence to achieve sustainable development and break-through in agricultural production, continuous concentration on technical advancement, input supply, credit and market supports are required. The Government is implementing various programmes to address the issues and constraints faced by the farmers to achieve the targeted growth in agriculture. The Government also primarily shoulders the major task of disseminating advanced technologies to 78.59 lakhs farm holdings through the departmental functionaries.

Tamil Nadu one of the leading rice growing states in India, has been cultivating rice from time immemorial as this State is endowed with all favourable climatic conditions suitable for rice growing. Frontline demonstrations are considered to be the most effective and useful extension activity to demonstrate the latest technologies developed at research stations to the farmers, in their own fields.

The principle of "seeing is believing" is operational in these demonstrations, as the farmers become easily convinced when they see the performance of new technologies in the fields of their neighboring farmers. Frontline demonstrations (FLDs) is formulated by the Indian Council of Agricultural Research (ICAR) and funded by the Government of India. Since the technology generators (scientists) are directly demonstrating and disseminating the technologies in farmers' fields, they are called "front-line" demonstrations. A comprehensive package consisting of new seed (variety/hybrid) and recommended cultivation and plant protection practices, etc., is demonstrated to farmers. Financial assistance is provided for critical inputs such as seed, fertilizer, weedicide, pesticide, etc. Organizing field days at an appropriate stage of the crop at strategic locations for a cluster of 20-30 demonstrations is an integral part of these demonstrations, which adds significantly to their effectiveness. These field days provide an on-the-spot opportunity for a large number of interested farmers to know the advantages of new technologies.

In last 6 years, about 7007 FLDs of 1 hectare each have been conducted benefitting 18318 rice farmers directly. About 2637 tonnes of seed of high yielding varieties/Hybrids was distributed to farmers through this programme. Spin off impacts of FLDs are many. The FLD programme helped many stakeholders to get first hand information about newly released varieties/hybrids that went into the value chains of Indian rice sector. Field demonstrations are a long term educational activity conducted in a systematic manner in farmers' fields to show worth of a new practice/technology. The impressive yield advantage obtained in the various ecosystems proves the fact that the FLDs conducted have been able to fulfill this objective. Overall, this approach has proved to be a very effective tool for the transfer of new technologies. Other rice growing countries can develop insights from the Indian experiences.

Rice is already a staple food for half of humanity, and the human population is set to grow to 9.5bn by 2050. Rice production therefore faces major challenges, to increase productivity, environmental sustainability, to address crop diseases and address micronutrient deficiencies, all within the context of a changing climate. Yet rice, like wheat, remains locked out of the two-decade-old biotechnology revolution. Largely because of political activism motivated by superstition and anti-science attitudes, there is no GM rice available to consumers anywhere in the world to this date. Projects to produce Golden Rice, nitrogen-efficient rice, insect and disease-resistant rice for example, utilize germplasm from a much wider variety of sources than is available through conventional breeding. Technological innovations in rice have made significant contributions to yield improvements. The impressive yield growth posted in past decades can be traced to increased use of modern inputs, products of previous investments in R and D–new rice varieties and crop management practices (Bordey, 2010). Led by PhilRice with its network of

57 members, local R and D activities have focused on four main areas: (i) breeding new rice varieties; (ii) improving crop management practices; (iii) developing appropriate farm machinery; and (iv) integrating rice farming with other agricultural activities. Biotechnology and conventional breeding techniques are used to develop new varieties that perform well under specific conditions irrigated areas, rainfed lowlands and uplands, cool and elevated areas, and saline-prone areas. Advances in rice research have typically veered towards increasing food production to feed a growing population. These efforts further need to be coordinated to combine high yield with superior quality and nutritious grains.

Rice plays a critical role in food security and poverty alleviation. The rice industry faces increasing challenges due to climatic, demographic, economic and social changes, shrinking arable land, and the rise of social media, among others. To overcome these challenges and ensure resiliency of the rice supply chain, building a collaborative and interdependent network among the governments, private sectors, non-governmental organizations, research institutions and the farmers is needed. Considerable attention has also been devoted to establishing voluntary sustainability standards for a number of high-value traded commodities. One example is the Sustainable Rice Platform (SRP), a multi-stakeholder initiative launched by UNEP in collaboration with IRRI to establish sustainability standards for rice. The coastal zones of the Asian mega deltas have largely missed out on the benefits of the Green Revolution. Productivity is hampered by too much water in the rainy season, too little water and/or salinity in the dry season, and cyclonic events. These problems will be exacerbated as a result of climate change. Yet there is tremendous potential to greatly increase productivity through the use of improved germplasm, cropping system intensification and diversification, and improved water management – especially drainage, but also by making better use of existing water resources for irrigation. Rice in less favorable environments often faces several abiotic stresses within the same season. Improved varieties need to incorporate tolerances to multiple stresses to cope with current conditions and future adversities caused by climate change. Good progress has been made over the past decades. In most cases, stress tolerant varieties need to be managed differently to maximize their supremacy in farmers' fields. Certain management practices that work in intensive systems could either have similar effects, no effect or negative effects in less favorable areas.

What is Ahead?

A number of other countries are expecting to begin or increase rice exports in the near future, such as Argentina, Brazil, Cambodia and Myanmar. In sub-Saharan Africa, where nearly half of the rice is imported, big opportunities exist to expand production, replace costly imports, and help offset malnutrition and poverty. Latin America, with its ample land and water resources, may ultimately become transformed into a major exporter of rice, thus helping to stabilize the global rice market. More rice exports from a larger number of countries will help buffer future trade against some of the causes of instability that have been described. However, world prices are likely to remain unstable, as production shocks occur or trade policies change in the major exporting countries. Climate change could also contribute to the instability in prices, depending on how it affects productivity in rice-producing countries.

Lingering pessimism about the reliability of international markets in the wake of the international price crisis of 2007–2008 may slow growth in the rice trade. This could change if proposals to reduce international price fluctuations are adopted.

Ultimately, poor rice consumers and net rice-deficit farmers in rice-importing countries generally benefit from trade because of lower prices and greater availability. For these hundreds of millions of households, as well as for rice producers in exporting countries, continued expansion of world trade in rice would enhance welfare and promote food security.

References

1. Canadian Agriculture Trade Statistics -2010.

2. Food staples Sufficiency Program -2011- 2016. 2012. Department of Agriculture Quezon City, Philippines. www.da.gov.ph.

3. Georges Giraud. 2013. The World Market of Fragrant Rice, Main Issues and Perspectives – A review. *International Food and Agribusiness Management Review*; Volume 16, Issue 2, 2013

4. http://ricecongress.com/2014.

5. http://www.rkmp.co.in

6. India – October Crop Review and 2014. Winter Crop Prospects, 2013. Informa Economics Survey-Based Crop Reporting Service.

7. Marc J Cohen. Food supply, factors affecting production, trade and access (Chapter 32)

8. The World Bank. "India: Priorities for Agriculture and Rural Development." The World Bank. 2011. Web. 7 Mar. 2011.

9. U.S. Commercial Service. "Doing Business in India: 2010 Country Commercial Guide for U.S. Companies." U.S. Commercial Service. 2010. Web. 2 Mar. 2011.

Chapter 2

Landraces of Rice (*Oryza sativa* L.) in Tamil Nadu: Their Conservation and Utilization

M. Subramanian

Former Director of Research,
TNAU, Coimbatore, Tamil Nadu
e-mail: drmsmanian@yahoo.com

Tamil Nadu one of the major rice growing states in India, is enriched with a lot of rice biodiversity. Rice, the most important food crop of the state is under cultivation from very longtime back, continuously from generation after generation. Farmers used to cultivate more than 2000 landraces/traditional rices which are one of the important genetic components of biodiversity. These landraces had a wide variations for many desirable traits like stature, tillering ability, panicle structure, grain colour, number and quality, aroma and medicinal properties besides, many of them were reported to be tolerant/resistant to abiotic and biotic stresses. Rice landraces with tolerance to salinity and alkalinity, submergence, flood and deep water conditions were also existed. These landraces of rice were utilized in breeding new rice varieties and also to improve and enrich with desirable traits of deficit varieties. In the beginning the plant breeding was initiated with pureline selection and many purified and improved versions of the already existing landraces of rice were released for large scale cultivation, subsequently the high yielding, tolerant/resistant and quality landraces were utilized to develop new high yielding modern rice varieties by hybridization. Unfortunately many of the landraces were extinct or verge of extinction due to the introduction of high yielding varieties and hybrids of rice and also due to various other reasons. However, realizing the value and importance of these indigenous genotypes, many of them were collected and conserved in some rice research stations of Tamil Nadu Agricultural University particularly, the Paddy Breeding Station, Coimbatore and Tamil Nadu Rice Research Institute, Aduthurai are maintaining more number of landraces under rice germplasm conservation to stop the loss of the valuable landraces and also for rice improvement work.

Biological diversity/Biodiversity is defined as the variability among bio organisms from all sources and also explains essentials of each species. Agricultural biodiversity is a portion of biodiversity that has undergone continuous selections and modifications over milliennia by human civilization to better serve their needs. Genetic biodiversity one of the three components (Species and Ecological) of biodiversity refers to the variety of plants, animals and micro organisms. Plant diversity is not only distributed over many countries in the world but also in India, which is one of the 17 mega biodiversity areas in the globe with enormous diversity in many flora and fauna (Mittermeier, 1988). Of the 4 per cent of 25000-30000 known edible plant species, only 150- 200 are used by human beings. Among them only 3 crop plants *viz.*, Rice, Wheat and Maize are important and playing very significant role in feeding the ever growing population of the world and provide 60 per cent of the calories and needed protein to the humans (FAO, 1999b).

Rice (*Oryza sativa* L.), the staple food crop for more than 50 per cent of the global human population, is next to wheat. It belongs to the family Poaceae (Gramineae) and tribe Oryzaea. The tribe Oryzaea consists of 12 genera including the genus Oryza, with specific differences among their traits.The genus Oryza in addition to 20 wild species has two cultivated species *viz.*, *Oryza sativa*, the Asian cultivated species with two ecotypes *i.e.*, Indica and Japonica and the African cultivated species *Oryza glaberrima* (Vaughan, 1994).

Unlike other crops, rice is gifted with enormous biodiversity, scattered in many parts of the world, particularly they are abundantly present in South East Asia and African countries. The components of the biodiversity comprising modern varieties like obsolete varieties, landraces, genetic stocks, breeding lines and wild rices, are the basis for the food security.

The landraces, the traditional varieties and the wild species are reported to be the genetic wealth of the country, because they are enriched with valuable gene system. Many have reported the existence of landraces of rice in India with varying numbers. Richharia and Govindasamy (1990) reported that a lot of evidences were there to show that India was having about two lakh varieties of rice.

Landraces of Rice

Landraces of rice, the most valuable component of genetic diversity, are viewed differently by various experts and were defined based on their origin, appearance and utility. They are local varieties of domesticated plant species, which are adapted to different natural and cultural environments in which they live and are the basis of resources upon which entire plant breeding depends. They are genetically heterogenous inspite of their autogamous nature (self pollination). Zeven (1998) reviewed the various definitions about landraces. In general It is opined that domestication of wild rices resulted in the development of land races from which the traditional varieties originated by evolution through natural selection or drift and local human involvement like crop cultivation, without adopting any breeding application over a long period of time. Therefore it is the intermediate form between wild rice, and traditional rice and serve as reservoir of genetic variation and biodiversity.

According to Harlan (1975), land races are variables and developed as a result of natural selection over millennia which also included human involvement. They are genetically diverse, with mixture of genotypes which are dynamic, recognizable, morphologically differ in their adaption to soil types, time of sowing, date of maturity, height, nutritive values, use, resistant to pests and diseases and other properties and farmers name for them. Landraces are highly balanced variables inequilibrium with both environment and pathogens and heritage from past generation of cultivars.

Zeven (1998) defined the autochthonous (indigenous) landrace is a variety with high capacity to tolerate biotic and abiotic stresses resulting in high stability and intermediate yield level under low input agriculture system. The term landrace/traditional varieties are some times used inter changeably. Landraces are grown from seed not have been systematically selected by seed companies/ developed by breeders. Landraces are referred to all those cultigens that are highly heterogenous but, with enough characteristics in common to permit them recognize as a group.

Genetically diverse landraces that are grown, are kept by the farmers especially in the vicinity of centre of diversity. Land farmers in many regions of the world growing land races for a very long period continuously because they are highly adaptable to local conditions and to variety of agro-ecological nitches, identifiable and usually has a local name. It lacks formal crop improvement and tolerant to biotic and abiotic stresses and is closely associated with the uses of the people who developed and continue to grow it (Vetelainen *et al.,* 2009).

Why Landraces are Important?

Important components of genetic biodiversity and most economically valuable and useful Adapted to specific local conditions and developed for regional uses. The ability of land races to withstand adverse conditions and exhibit the important traits is better than HYV (Rajukannu *et al.,* 2009). High and wider genetic variability of old land races help to have genetic flexibility to adapt local field conditions, changing environments, farming practices and have specific uses for human and animal consumption (Deepakkumar Rijal, 2010). The farmers and the tribals who take care of the landraces of rice are continuously cultivating them without break because, small and marginal farmers and tribes are not affordable to incur high cost and to take up the cultivation of modern high yielding rice varieties and hybrids. Highly associated with the religious and cultural functions. Many races are having very good aroma and flavour and other special characteristic features. Many landraces of rice have been identified for their good medicinal properties to treat various ailments of both human beings and domestic animals. Grains of Landraces/traditional varieties fetch higher market price. The straw of land races are mostly long (Tall stature), can be used for thatching huts and also has good feed value, therefore fetches higher market price. When water is adequate and naturally N fixing organisms exist, no synthetic fertilizers required to produce 2 tons/ha by a local rice variety (Swaminathan, 1984). Many landraces are highly drought tolerant, even with minimum rainfall and they perform well. The deepwater rices may grow 18'above the water level under low lands in the seasonal waters. When yield is calculated as per directions of grains/unit input, most of the local races have better values than HYV (Deb1995, 2000). In a simple cultivation technique the transplanted land race gives good seed ratio when compared to HYV grown with high technique of Green revolution. Useful for the preparation of many eatables like snacks, sweets, Payasam, Rice cakes, Flakes, Puffed rice, many delicacies and special food preparations suchas Pongal, Biriyani, Polov, Fried rice etc., In addition some rice landraces are used to prepare beer and wine.

Rice Landraces of Tamil Nadu

Tamil Nadu, the 11th largest and one of the important southern states is located in the south ern most part of the peninsular India. It is bordered by Andhra Pradesh in the north, Kerala in the west, Indian ocean in the south and Bay of Bengal in the east. It is spread in 1303372 sq km as geographical area comes under subtropical climatic region. It is having a mild climatic conditions receives an annual rainfall ranging from 635 to 1195 mm with an average of 925mm. The north east monsoon records a high rainfall (47.42 per cent) followed by southwest monsoon (33.25 per cent), summer (14.75 per cent) and winter (4.56 per cent) seasons. The state has recorded minimum and maximum temperatures, ranging from 18°C to 43°C.The high temperature is usually recorded during April-June while the temperature goes down during November-February (Season crop reports of 2009-2010, Govt; Tamil Nadu).

Tamil Nadu is a very old and ancient rice growing state, where more than 2000 landraces were reported to be under cultivation. Almost all the districts of Tamil Nadu had rice cultivation with landraces of rice, particularly, they were more in number in Thanjavur, Trichy, unbifuricated NorthArcot (Vellor and Thiruvannamalai), and Chengalpattu (Thiruvallur, Kanchipuram), and unfiburcated Ramanathapuram (Sivaganga, Virudhunagar) and Thirunelveli, and were under cultivation for a very long time (Sathya 2014). In addition Kanyakumari, Pudukottai, Salem, Thoothukudi and Madurai districts had landraces/traditional varieties cultivated under diiferent ecosystems and environments (Subramanian *et al.*, 2004 and 2011) (Table 2.1).

Genetic Diversity of Landraces of Rice

The interaction between farmers' needs and environment has been the huge heritage of rice genetic diversity (Myers, 1994), which has been estimated to be more than 600 varieties of *Oryza sativa* in Tamil Nadu adapted to grow for past several years in this state (Subramaniyan, 2011). The actual number of landraces rice existed in Tamil Nadu was not predicted correctly, however, Sathya (2014), reported about 400 landraces existed in Cauvery delta (Wet system) alone during the Kingdom era. Ramasamy (1972) has reported that Ramanathapuram (Dry system) alone had about 100 landraces.

Table 2.1: The Landraces of Rice in Tamil Nadu

Achchumurithan, Adhivaragan, Adhukan (Adukkan), Adimodan, Aesadaisamba, Alpal, Althersamba, Amarathansamba, Amirthamsamba, Anaikomban, Anaikombansamba, Anjanam, Anjangumkumbamalai, Annamazhagisamba, Annasamba, Annathani, Annuruvi, Araisamba, Araisembalai, AraumKuruvai, Araunsamba, Aravankuruvai, Archunan, Arcotkichili, Arcotponni, Arikirai (Arikiruvai, Aruvikirvai), ArisambaAriyanayagam, Ariyannel, Arumbosasamba, Arunjyothi, Arupathamkodai, Arupathamkuruvai, Aruphathamsamba, Aruphathamvattam, Arupathamvellai, Arunthathi, Aruvankuruvai, Arvumsama, Aryan (Aryannel), Aryanred, Athura, Athursamba, Avasarasamba, Ayyansamba, Azhagiamanavalan, Azhagiavanan, Azhagusamba, Balanthattaravellai, Bangalurkar, Bangaruthegalu, Barkadukkan, Bayyakundan, Benhipedi, Boombalai, Boonsamba, Boronvellai, Buththanel, Buththavari, Charkarnel, Chelluvali, Chengalpattusirumani, Chenkar, Chenkuruvai, Chennel, Chenthazhai, Chenthee, Chettisamba, Chettivali, Chinnaadukkunel, Chinnasmaba, Chinnagam, Chinnamanavari, Chinnaponni, Chinnapunchai, Chinnasamba, Chinnasivappunel, Chinnavadansamba, Chirunalairayan, Chinthamani, Chitharyan, Chithraikali, Chithraikar, Chithraivannan, Chittensamba, Chomala, Cinnagam, Cochinsamba, Coimbatoresamba, Dhonanellu, Edavakka, EerukkusambaEluppaisamba, Erapallisamba, Eravapandi, Garudansamba, Ghandhasala, Gobianaikomban, Godamanisamba, Godhavarisamba, Godomaisamba, Gopikar, Gopisadai samba, Gowdalu, Gowri, Gundansamba, Idaikannan, Ilanthiraikondan, Illuppapoosamba (Iluppaisamba), Irangalmeetan, Iravaipandi, Jeeragasamba (Seeragasamba), Jeerkudai (Jirkudai), Jiljilvaiyakundan, Kadaikazhuthan, Kadukkan, Kadukkansamba, Kaiveraisamba, Kakkankuruva (Kakkakkuruva), Kakkarathan, Kalappukuruvai, Kalarkar, Kalarankuruvai, Kalarpalai, Kalarsamba, Kalavai, Kalingararayan, Kaliyansamba, Kallamsamba (Kallansamba), Kallanthattaravellai, Kallimadayan, Kallimandayan, Kallundi (Kallundai), Kallundaisaradi, Kallurundaikar (Kallurundai, Kalurundi), Kallurundiayan, Kalmanavari, Kalrankuruvai, Kalvidhaisamba, Kalyansamba (Kalyan), Kambansamba, Kanakattai, Kanchiinamkondan, Kancinamkondan, Kandasala (Kandasali, Kandasal), Kaniyalan, Kangaru (Kongaru), Kannadikoothan, Kannansamba, Kappakar (Kappikar, Kappar), Kappasamba, Kappansamba, Kar (Karpaddy), Karaamjoori, Kararisi, Karimundu, Karimuntakan, Karnel, Karpurapalai, Karsamba (White), Karsamba (Red), Karsivappu, Karthigaisamba, Karthivannan, Karudansamba, Karumbukali, Karumkalsmba, Karumpuzhuthi, Karunchoorai, Karunellu, Karungkuruvai, Karunjeeragasamba, Karunsamba, Karuppeynel, Karuppugowni, Karuppunel (Karinel), Karuppuputtu, Karuthamanavari, Karuthakar, Karuhasiradi, Karuthasirusali, Karuvalli, Karuvati, Karuvankuruva, Katchakoombalai (Katchakambalai), Kathalivazhai, Kathgaiisamba, Kathasamba, Kathurisamba, Kathurivanan, Katiyanam, KattaiAraisembalai, Kattaikulazhan, Kattaimundan, Kattaisamba, Kattaivalan, Kattaivellai, Kattalaguvannan, Kattanelllu, Kattanurnel (Kottanur), Kattaransamba, Kattikar, Kattimosanam, Kattisamba (Kattasamba, Katta, Kattusamba), Kattiyanam, KattuKayama, Kattukuthalam (Kattukuthalai), Kattuponni, Kattuvallai (Kattuvellai), Kattuvanam, Kattuvaniyam, Kattuyanam, Kavivarisamba, Kavuni (kavuninel), Keeraisamba, Keerisamba, Keralagandhsala, Kesarigandhsala, Kichilisamba (Kitchadisamba), Kirijasamba, Kochisamba, Kodai, Kodaikannan, Kodaikazhuthan, Kodaikuruvai, Kodaisamba, Kodaivilayan, Kolamthattaravellai, Kolavalai (Koolavalai), Kollansamba, Kollian, Kar, Komban (Kombansamba), Kongadumuthirnarisendalarangan, Konakuruvai, Konamani, Kondalarigiri, Kondalanilamerun, Kondan, KongKongarumuthirnarkendalangaran, Koodainel, Koolavazhai, Koombalai (Koomvazhai), Koothan, Koraisamba, Korangusamba (Kurungusamba), Kotanellu, Kothamallisamba, Kothandan, Kothumaisamba, Kottaramsmaba, Kottikkar, Kouninel, Kuchivedichan, Kudaivali, Kudavazhai (Kuzhavalai), Kudavaliyan, Kudir, Kudhiraivalan (Kudhiraivalsamba), Kudhiraivali, Kudhiraivalsirumani, Kuduvalai (Kudavalai, Kudaivazhai), Kulakkuruvai, Kulvazhai, Kuliyadisamba (Kulivedichan, Kuliyadichan, Kuzhiadichan), Kullankar (Kullakar), Kullavellai, Kuluvathivalayan, Kuminijan, Kundusamba, Kundali, Kudrimanisamba, Kungumapalai, Kunjinamkondan, Kunthali, (Kundhalai), Kuppaisamba, Kurakannan, KuravathivalayanKurukot (Kalyani), Kuruva (Kuruvai, Kuruvainellu, Khuruvai), Kuruvaikalangium, Kuruvaikalayan, Kuruvikar, Kuruvaikillai, Kuruvaikiliyan, Kuruvaisornavalai, Kuthippan, Kuthirupu, Kutralam, Kuvalai, Kuzhavazhainel, Lakshmikajal, Machumurithan, Madhumilangi, Madhumuzhungi, Madhuraivanan, Madhuriveli, Mahalaksmi, Mahata, Maisamba, Malaikichadi, Malaimundan, Malainellu, Malayalasamba, Malligai (Malligaisamba), ManakattaiManakathai), Manavari, Manalvari, Mangalpuramnel, Mangamaikathan (, Mangamarkathan), Mangalsamba, Mangamsamba, Manikkamalai, Manikkasamba, Manilaponni, Manisamba, Manjalponni, Manjalsara, Manajalsirusali, Manjumurinathan, Mannuvaliyan, Manvilayan, Mappillaisamba, Maranel, Marthondi, Marudhi, Mashanam (Moshanam), Masikar (Maasikar),

Contd...

Table 2.1–*Contd...*

Mathimuni, Mattai, Mattaikar, Mattaikuruvai, Maylapur, Meenampur (Meenamburi), Meenampini, Menaminikki, Milagusamba, Mikkuruvai, Minjammakuthiraivanan, Misamba, Mohinisamba, Molagoluklu, Mookannel, Mookansamba, Moolai, Moongilsamba, Moonginel, Mookan, Morungankar, Mosam, Moshanam, Mottaikar (Mottakar), Mottaikaruppan, (Mottakuruvai), Mottanellu, Mozhikaruppan, Mudgo, Mullampanchan, Mundankuruvai, Mundamaranellu, Mungilsamba, Murungangarnel, Murappan, Muthirnarikondakkaran, Muthumalai, Muthumani, Muthusamba, Muthuvellai, Muthuvilangi, Muttakar, Muttampanjan, Muzhuvellai, Mysoremalligai, Nagarayan, Nallakonamani, Nallamanisamba, Naraiyan, Narkuthalan, Navara, Navarai, Navaraikali, Navarasakali, Nedumookan, Neelansamba, (Neelasamba) Nelethirsamba, Nellursamba, Nensambakuruvai, Nirkamindan, Nootripathu (110), Norungan, Norungankar, Norungansamba, Oazharkatrazhai, Omathikiradhi, Omadhikiradhithalavadi, Onjakarpurapalai, Ondaraikitchali, (Ondaraikichadi), Ondaraisamba (1½samba), Osarakuthalai (Osuvakkuthali), Othakomban, Othukitchali, Ottadai, Ottadan, Ottadayan, Ottansamba, Ottu, Ottuchandi, Ottumanavari, Ovarkondan, Pachaiperumal (Pitchaiperumal), Pagunapalai, Palansamba, Palkatch, Palkudavazhai, Pallanthattaravellai, Palmansalf, Palthondi, Palvadiyumsamba, Pamanisamba, Panaimarathusamba, Panaimoran, Panamkuruvai, Panamoori, Panangattukodavalai (Panangkattukuzhavazhai), Pandisamba, Panneersamba, Parakadukkan, Periyasali, Pathrakali, Pattanamsamba, Pattaraikar, Pattarpisini, Pattinamkathan, Pavaibham, Pavalasamba (Pavalam), Pavithramsamba, Pendasinivallai, Periakichili, Periasamba, Periyachandikar, Periyakitchalisamba, Periyavari, Perunkar (Periakar), Perumbalamsamba (Perumbalam), Perumsamba (Periyasamba), Perunellu, Perunkar, Perunthanduvellaisamba, Peruvellai, Peruvellaimokkan, Peruvellaisamba, Pillansamba, Pisini, Pitchakaruppan, Pitchavari, Pondisamba, Ponkombisamba, Ponmanisamba, Ponnariyan, Ponnayakan, Poombalai, Poonchali, Poonkar (Punkar), Poonsamba, Poovansamba, Poravellai, Poranvellai, Porapalai, Porsali, Powkombisamba, Pudupttisamba, Pudhuvithu, Punugusamba (Pongusamba), Puzhthikkal, Puzhuthikar, Puzhuthiperetti, Puzhthiperattikar, Puzhuthisamba, Rajarajavanan, Rajameni, Rajayoham, Ramabanam, Ramakuruvakar, Rangoonsamba Rascadam, Ravanan, Redsirumani, Redottadan, Sadakar (Sadaikar), Sadari, Sadaisamba, Sadayali, Salemsamba, Salemsanna, Salemvelim, Samba (Sambanellu), Sambamosanam (Sambamosam), Sandikar, Sannakali, Sannasamba, Sannasornavari, Sannavellai, Saradi, Sarapuli (Sarapalli), Sarapallisamba, Sarkarnel, Seenakali, Seenellu, Seeragasamba (jeeragasamba), Seetabogham, Seethavalli, Seevanusamba (Seevansamba), Sembalai (Chempalai), Sembilipanni, Sempavalasamba, Sempulikuruvai, Sempuliyan, Sempulipriyan, Sempulipuramsamba, Sempuliseenkanni, Sendalankaran, Senkaladi, Senkar, Senkuruvai, Sennel, Sensamba, Senthadi, Senthazhai, Senthinayagam, Settuvalayan, Shalikar, Shenmolagai, Sigappuguzhiadichan, Singan, Singarakallurivanam, Singarathurivanan, Sinnasamba, Sinnasivappu, Siraimeetan, Sirumani, Sirumanikuminchan, Sirusamba, Sithundikar, Sivappuchitraikar, Sivappujeergasamba, Sivappukar, Sivappukarsamba, Sivappukavuni, Sivappukud (t) avazhai, Sivappukuruvaikalyan, Sivappukuruvaikar (Sivapukar), Sivappukuzhadichan, Sivappunel, Sivappuottadan, Sivappuputtu, Sivappuseeragasamba, Sivappusirumani (Sivappujermani) Sivappusornavari, Sorikurumbai (Sorakurumbai), Sornamashuri, Sornavalai (Sornavali), Sornavari, Sornavarikuruvai, Soolarkuruvai, Soorakuruvai, ((Surakuruvai, Surankuruvai) Soornumkuruvai, Sugadoss, Sumalai, Thalavadi, Thanganel, Thangasamba, Thangathakaikolai, Thanjavurvadansamba, Thappakaransamba, Thattansamba, Thattaravellai (Thattarapillai), Thekkan, Thellavadansamba, Therkathikar, Thidakal, Thillaikoothan, Thillainayagam, Thinni, Thiruchengodusamba, Thirumangaialvan, Thirumangaiazhagan, Thirunelvelinathan, Thirunelvelivanan, Thiruthuraipoondakar, Thiruvarangan, Thiruvengadan, Thuyamalli, Tinnyankomban, Thodavaliyan, Thogaisamba, Thondi, Thooyala, Thooyamalli (Thuyamalli), Thotasamba, Thulukkasamba, Thulunadan, Thumbaipasi, Thungara, Ulaguvaiyakundan, Uppumilagi, Urundaikar, Uthirikar, Uvarkondan, Uvarmeetan, Uyya kondan) VadaArcotvadansamba, Vadakkathikar, Vadakkathisamba, Vadansamba, Vaikaraisamba, Vaikunda, Vaiyakundan (Baiyakundan), Valaichitraikar, Valaithalaisamba, Valamsanna, Valan, Valaannel, Valiyan, Valiyansamba, Valkaruppan, Vallalnellu, Vellapuzhuthi, Vallarakkan, Vallikar, Valsamba, Vanvellai, Valsaramundan (Vasarumundan), Vanakkar, Vanannel, Vanginarayanan, Vangisamba, Vanguvellai, Vangvarabanpansamba, Varagampansamba, Varakkalnel, Varalan, Varalisamba, Varamilagi (Varagumikagi), Varappukudanchan, Vardansamba, Varigarudansamba, Varisamba, Varisraimeetan, Varisuriyan, Vasanaijeeragasamba, Vasanaikayama, Vasanaillathakayama, Vazhai, Vazhakondan, Vedakkallidanta, Vehirisamba, Veedhivadangan,

Table 2.1–*Contd...*

Veeradangan, Veeramarthandan, Veersadangan, Veerasannvadlu, Velamsamba, Velanchanel, Velari, Velchi, Vellai (Vallai), Vellaichitraikar, Vellaigouni, Vellaigundu, Vellaikar, Vellaikarsamba, Vellaikaryan, Vellaikattai, Vellaikundusamba, Vellaikudavazhai, Vellakuruvai, Vellaikuruvaikar, Vellaimarakkattai, Vellaimautta, Vellaimudangan, Vellaimuthan, Vellainel, Vellaiottadan, Vellaipatharai, Vellaipoongar, Vellaiputhan, Vellaiputtu, Vellaisamba, Vellaisinnamnai, Vellaisirumani, Vellaisirusalai, Vellaisornavari, Vellaithandu, Vellaithattai, Vellarakkan, Vellayan, Velumbalai, (Velumpalai), Velumphazha, Veluthankudir, Vendakulavazhai, Venginel, Vengisamba, Venguvellai, Vennel, Verakudavazhai, Vethrisamba, Villudrisamba, Vilupuramsamba, Viralisamba, Whiteputtu, Whitesirumani, Yanaikavuni, Yeerasannavadlu

Ravisankar and Selvam, 1996; Vijayalakshmi and Nambi, 1997; Chellam, 2010; Subramanian *et al.*, 2010 and 2011; Anurudh K. Singh, 2013; Parimalathirumurugan, 2014; Wickipedia; Sathy, 2013 and 2014 and Personal communications from Dept. of Agriculture and Research Stations of TNAU.

Genetic diversity of Tamil Nadu rice landraces is found with a wide range of variations for their morphological characters as well as high ecological and physiological adaptability to grow in varied agro ecosystems like uplands, wetlands hilly tracts, valleys, flooded areas, poor, toxic deficient and problem soils. This crop has been cultivated under irrigated, rainfed, shallow land flooded conditions, and deepwater and mountain areas, in addition, this food crop was grown in areas with coastal and inland salinity problems.

Landraces of rice with different maturity periods [The short (About 60 days), medium(120-140days) long duration(141-160 days) and very long duration (> 160 days)] were under cultivation and were having tremendous morphological variability for traits such as stature of the plant(very tall, tall, medium tall, medium dwarf and dwarf) tillering ability, leaf blade colour (pale green, green, dark green, purple margin, purple blotch, purple) colour of leaf sheath(green, light purple and purple) panicle type (open, intermediate, compact), awn with varied lengths (awnless, short, partially, awned, long fully awned), apiculus colour (white, brown, purple, purple tip), colour of lemma and palea (straw, golden furrows, brown spot, brown furrows, brown, light purple, purple spots, purple furrows, purple, black), kernel colour (white, light yellow, light red, red and deep red), grain pubescence(long velvety hair), panicle no/hill, panicle length, panicle weight, number of filled grains, 1000 grain weight, kernel length, breadth, length/breadth ratio, fine grain, course grain, long slender, medium slender, short slender, long bold, medium bold, short bold, grain textures like transluscent and opaque, threshability, hulling and milling recovery cooking qualities, protein, lysine and fat content, aromatic (mild to strong), non aromatic and medicinal properties. The landraces of rice are also possessing resistance/tolerance to biotic and abiotic stresses like pests and diseases and drought, submergence, flooded, deepwater conditions (4m) and various soil problems.

Nutritional and Medicinal Landraces of Rice

Tamil Nadu has documented many landraces/traditional varieties with high nutritional values and medicinal properties. Some of the landraces of Tamil Nadu with medicinal properties are mentioned here.

Annamilagei, Annuruvi, Arcotkichili, Arubamkuruvai, Arupathamkuruvai, Eirukusamba, Iluppaipoosamba, Jeeragasamba, Kadaisamba, Kaiveraisamba, Kalarsamba, Kallundai, Kambanchamba, Kararusi, Karunchamba, Karungkuruvai, Kathanellu, Kattisamba, Kavuni (Black) Kitchadisamba, Kodaisamba, Koombalai, Kudaisamba, Kundusamba, Kundrimanisamba, Kurunchamba, Kuliadichan, Kundusamba, Madhumizhangi, Malligaisamba, Manisamba, Mappilaisamba, Milagusamba, Neelansamba, Ottadayan, Palkhulavazhai, Panisamba, Pichavari, Poongar, Poovansamba, Rajabogham, Sivappukudavazhai, Sivappukuruvaikar, Sivappusamba, Vellaisamba, Veeradanganetc (Sathya 2013, Chellam2010 and Saravanan 2012) and their uses are presented in Table 2.2.

Table 2.2: Medicinal Landraces of Tamil Nadu and their Uses

Rice Landrace	Medicinal Property
Annamazhugi	Sweety, cures many Pitha diseases, and controls body temperature
Arcot kitchali	Easily digestible, good for feeding mothers and milc h animals, induces more milk secreation
Arupathamkuruvai	Good for sick patients
Eerkusamba	Good taste and induces sexual feelings
Iluppapoosamba	Produces more pitha, enhances body temperature and related diseases
Jeeraga samba	Cures diseaes of Pitha, Vadha and Kaba, induces apitite, easily digestible, keeps body healthy and bright, and increases semen count
Kadaisamba	Cures sexual diseases and increases spermcount
Kaiveraisamba	Increases body strength but raises Pitha
Kalansamba	Develops strong firmness and cures pitha diseaes
Kallundaiarisi	Increases body strength
Kararisi	Develops fatigue, increases body weight and Strength
Karunchamba	Sweet taste, reduces Pitha, Vadha and Kaba develops body colour and brightness, reduces irritation and increases sperm count
Karungkuruvai	Used for preparing Kadi, a base for Sitha med icines, cures leprosy with severe wounds, serves as an antidote, increases sexual feelings, and also cures flariasis
Kattisamba	Reduces blood sugar in diabetic patient
Kavuni (black)	Tasty, increases body strength and helps easy delivery
Kodaisamba	Reduces gastric pains
Kondalanilamerun	Reduces sugar in the blood
Koombalai	Reduces pain during delivery of child
Kundrimanisamba	Increases body strength, cures Vadha diseases and increases semen count
Kundusamba	Cures diabetics but develops fatigue due o heat and karappan disease
Kurunchamba	Good for curing pricking Vadha disease
Kuzhiadichan	If the mother, after delivering a female child consumes this rice she gets increased milk secreation

Contd...

Table 2.2–*Contd...*

Rice Landrace	Medicinal Property
Malligaisamba	Cures diseases like Karappan, Pramegam, eya irr itation due to heat and increases sensitivity of tongue, body strength, and firmness of the body
Manakathai	Reduces Shivisham
Manisamba	Changes weak bodied children to strong strength
Mappillaisamba	Gives very strong
Milagu samba	Increases body strength and cures Vadha diseases
Nallamanisamba	Reduces sugar
Navarakakali	Fermenting swellings and wounded parts
Ottadayan old	Young people if consume this rice they will be stronger even after becoming
Panichamba	Good for diabetics
Poongar	Good for pregnant women
Sivappu kudavazhai	Reduces sugar in the blood of diabetic patient
Sivappukuruvikar	When cold rice is eaten, apitite (feeling) delayed
Sivappu samba	Cures Pitha, Vadha, and Kaba diseases, good for bronchitis, caugh, irritation, and purifies blood, helps for free urination, increases spermcount, protects eyes, gives strength to body, promotes heart function, keeps toungue wet, removes throat problems, wounds, thus, helps for good, and clear speach, removes toxic substances and reduces knee pain
Vellaisamba	Sweety, removes Pitha problems, promotes body growth, removes fatigue and induces vigour

Source: Chellam 2010 and Saravanan 2012.

Aromatic (Scented) Landraces of Rice

Aromatic (Scented) rice is a special and unique diversity of rice, very much attracted by different countries of the world and different sectors of people suchas farmers, traders, consumers and scientists. Some of the aromatic rices possess special quality like long slender fine grains coupled with pleasant aroma (Ahuja *et al.,* 1995, Khatana 2009).All the rice growing states in India is having many aromatic rices of its own and the number varies, but Tamil Nadu is reported to have only a few indigenous aromatic rices like Punugusamba, Rascadam, Sugadass, Rasanam, and Jeeragasamba (Sathya 2014) but now Jeeragasamba, alone is under cultivation for a very long time because it is special variety utilized for many family functions and also for offering to deities during festival days. This rice is generally consumed with meat during special functions. Since it contains B sitosterol which does not allow the cholesterol to accumulate but sent outside through stools (Madhanmohan *et al.,* 2013). The area under this variety has been reduced considerably however, it is still under cultivation in batches in a smaller area.

Landraces with Tolerance/Resistance to Abiotic and Biotic Stresses

Among the many desirable traits of landraces of rice, tolerance/resistance to abiotic and biotic stresses is very important and the rice accessions identified as resistant/tolerant were utilized for imparting the resistant genes to the susceptible modern cultivars, which is not only saved the crops from these stresses but also reduced the expenditure on plant protection. Many rice landraces of Tamil Nadu were screened for their resistance/tolerance in the Rice Research stations in Tamil Nadu and also National and International Research Institutes and utilized for resistance breeding and are presented hereunder.

Tolerance to Abiotic stresses

Drought Tolerant Landraces of Rice

Aryan, Chitraikar, Kallamodan, Kallurundaikar, Kallurundaisamba, Karumpuzhthi, Karuthamodan, Kattismba, Kattuvanam, Kuzhiadichan, Kullangkar, Karumpuzhthi, Kuruvaikalangium, Mettasannavari, Nootripathu (110), Norungan, Norungankar, Norungansamba, Poongar, Puzhthikar, Puzhthipretti, Puzhthisamba, Sivappukuruvaikar, Soorankuruvai, Vaiyakundan, Varappukudaanchan, Vellaichitrakar.

Tolerance to Water Stagnation, Submergence, Flood, and Deepwater

Karumpuzhthi, Kattuvana, Kudavazhai, Neelamsamba, Palkudavazhai, Sivappukudavazhai, Soorakuruvai Thiruthuraipoondikar, Vaiakundan and Varappukudnchan.

Tolerance to Salinity and Alkalinity

Arupathamkuruvai, Kalakar, Kalarpalai, Kalarsamba, Kallundi, Karumundagan, Kuthirnellu, Kuthirvithu, Panaimarathusamba, Osuvakuthali, Sivappukulavazhai, Surankuruvai and Uvarkondan.

Resistance to Biotic Stresses

Stemborer

Arunukuruvai, Koolavazhai, Sadaisamba, Sirunmani, Sirunsamba and Sornavazhai.

Brown Plant Hopper (BPH)

Anaikomban, Sadaisamba, Sinnasivappu and Sivappusirubsamba.

Uses of Landraces of Rices

The tribal people and the farmers in Tamil Nadu, who own and cultivate these landraces are fully utilizing them for very many purposes particularly for their every family and social activities. The processed and cleaned rice immediately after harvest are utilized for many family and social functions, festivals, religious ceremonies, rituals, marriages and death functions. They need different rice races for different purposes and for each function a specific variety of landrace of rice is utilized. Therefore each and every activity and cultural practice of farmers and tribal's are highly associated with keeping the biodiversity of rice safe. Particularly the utility of aromatic rices, flood tolerant rices (4-7mt depth) drought tolerant rices, tall varieties with high feed value and thatching huts, were given importance in maintenance and conservation. The aromatic and quality rice landraces had a very good demand and market values compared to modern high yielding rices.

In addition, the landraces are very much useful for many other purposes, besides used as food after cooking, it is utilized to prepare Idli, Dosa, Oothappam, Idiyapapam, Payasam, Palpani yaram, Puttu, Murukku, Seedai, Kolukkattai, Sweets, Sweetballs, Thenkuzhal, Athirasam, Delicacies, Riceflakes, Cakes, Puffed rice, (made out of course grains)varied eatables and bread (from both red and white rices) Noodles and Flattened rice, because when they are prepared from a specific landrace that gives the expected taste. Edible oil is also extracted from the land races. By utilizing the traditional rices alcohol, beer, wines, medicines etc; are also prepared and having very good demand in market. It is reported that food materials prepared from the landraces of rice have very good preference particularly food from red rices are tastier and swell and expand well and gives good volume of food compared to any modern high yielding rice varieties. Similarly eatable and food materials prepared from aromatic rices are also very much preferred by the traders and consumers.

The farmers and tribal's who cultivate landraces are processing harvested paddy manually, where the rice bran is not usually removed from the kernel, therefore the cooked rice from them are highly nutritious with high amount of protein, fats vitamins and minerals. Besides many rice landraces have been identified with medicinal properties and utilized for curing many human and animal ailments.

The rice land races are generally grow tall with stiff straw, therefore the straw are very much useful to thatch the houses and feed the cattle. The hulling and milling qualities are good in traditional rices and fetches good market prices.

Regarding cultivation aspects, short duration landrace varieties are very much suitable for raising mixed cropping with pulses, millets and oil seeds. The farmers of Ramanathapuram district of Tamil Nadu used to raise redgram along with local rice variety of rice and get additional income from rainfed rice cultivation. Broadcasting with seeds of two to three duration landraces is also common practice in tribal areas so that even one variety fails due to either inadequate or failure of monsoon and other varieties will give yield with the help of subsequent rains. Simillar practice was also under practice in Kanyakumari district of Tamil Nadu where, the farmers mix 5 groups of rice (Anginam, five types) with different durations and broadcast, the farmers are assured of getting full harvest or any one or two, or three or four, depending upon the rainfall received during crop growth period.

Genetic Erosion of Landraces of Rice

This wide genetic variations had their own deterioration due to very many reasons with the result many landraces have been disappeared and many more are in verge of extinction and become endangered due to Genetic erosion/Genetic vulnerability/Genetic wipeout. This phe nonmenon not only eroded the valuable and useful landraces but also damaged diversified cultures and experiences and promoted regional disparity (Subujkumar Choudhary).

What is Genetic Erosion?

Genetic erosion is the result of negative influence of many factors, which played a very critical role in reducing both loss of species, primitive races and reduction of cultivated varieties. The landraces were gradually replaced in the agriculture with the newer and more productive crop varieties and many other environments and human interferences like, 1. Pressure on land, 2. Social destruction, 3. Environments, 4. High yielding varieties and hybrids. 5. Introduction of alien species, 6. Use of few cultures highly related (Narrow sense), 7. Reliance of monocropping, 8. Basic stresses like pests and diseases and weeds, 9. Input cost, 10. Financial support for promoting modern varieties and hybrids, 11. Degeneration of materials during storage and tissue culture, 12. Small samples and pollinating systems etc.

Among the aforesaid factors, introduction of high yielding varieties and hybrids and the efforts taken to promote them in the name of Green revolution was the most important which played a very critical role in eliminating many land races in a shorter period of time. The introduction of HYV also had very dominant impact on production and productivity of rice, as they responded to higher input management as against the indigenous rice genotypes which had very poor response to these input management, therefore the replacement of landraces by HYV though gradual but in many places it was complete. During 1960 ie; the green revolution period introduction of IR8 and ADT 27 (Indica/Japonica) the high yielding rice varieties in this state replaced many traditional varieties like Kurangusamba, Kichadisamba and even made them to disappear, inaddition Kappakar a flood and drought tolerant variety from Madurai district and Sambamosanum a variety from Trichy have also been lost.

The impact of introduction of HYV through green revolution was very great and significant in eroding TV/LR which ranged from 10-100 per cent in Tamil Nadu (Personal information from Agri. Dept. Tamil Nadu).

In addition, environment and climatic changes and fluctuations also had significant effect in elimination of landraces. In Tamil Nadu, most of the landraces were cultivated either as rainfed or semidrycrops in Kanyakumari, Thirunelveli, Thoothukudi, Ramanathapuram, Sivaganga, Pudukottai, Thiruvallur, Salem, Kancheepuram, Madurai, Cuddalore, Trichy, Thanjavur, Nagapattinam, and Thiruvarur districts and their success is mainly depended on the amount and frequency of rainfall received during the crop growth period, whenever the rainfall is inadequate or failed completely or failed during reproductive stage, the rice cultivation also failed or the cropping pattern is changed, which paved the way for the elimination of landraces.

The special varieties like Kudiraivali (Fine grain) Putturices (Red and white types (used to prepare special eatables) Uvarkondan and Panaimarathusamba (Saline and Alkaline tolerant), white and sivappu Ottadan, (Very long duration *i.e.,*

220 days), Thiruthuraipoondikar and Kattuvanam (Deep water rice) need worthy to mention of their disappearance due to introduction of HYV.

Conservation of Rice Landraces

Exploration, collection and conservation of rice germplasm were very important to protect the valuable rice genotypes from erosion and also for crop improvement programmes. In the earlier period many and different rice landraces were conserved by the tribes and farmers continuously without any loss because of their own specificity and importance besides many land races were essential and needed for different purposes. Choosing the rice races for different sowing seasons, agro climatic conditions, against natural hazards and adverse conditions, varied rices for different functions, ceremonies, festivals, marriages, death funeral, social activities etc; was very important, which influenced growers to maintain and conserve them without any loss for their continuous uses. Mostly collection, processing, and storing the seed materials were looked after by the farm women. Tribal women in Tamil Nadu possess traditional ecological knowledge of plant species used for food and medicines, which played a significant role with the conservation and sustainable use of agricultural biodiversity Women in Javadi Hills, Chitteri Hills and Anamalai Hills were responsible for collecting storing viability of seeds of landraces.

The value of the genetic diversity, specifically the importance of rice landraces is realized only when they become endangered and a lot of landraces disappeared or gone out of cultivation from the field, natural hazards and human interference have been attributed as reasons for such loss of landraces (Frankel and Hawkers, 1975).

The genetic erosion was even to the tune of 100 per cent in many places by the introduction of a few HYV and they were all ever lost. They cannot be replaced or created because they are the natural gift to the Universe to help the living organisms including human beings. Ravisankar and Selvam (1996) and many others recommended active measures to conserve traditional varieties to prevent or slow the introduction of Modern varieties. This condition made the Government and Scientists to think twice about the importance of landraces, which forced them to plan and speedup the exploration effectively to collect, conserve and utilize the landraces of rice. The exploration commenced during 60s in India with a view to first study their origin and variability, utilize them to breed new varieties, to improve the rice genotypes which are deficit for some needed and desirable characteristic features and also to stop further loss of these important and valuable genetic materials.

In Tamil Nadu, collection and conservation of rice germplasm was started immediately after the establishment of Paddy Experiment Station at Coimbatore during 1912, which was the National Research Centre for rice research at that time. It is reported that Dr.K.Ramaih the then crop specialist in Paddy Breeding Station Coimbatore took away more than 2000 rice germplasm to CRRI Cuttack when he was appointed as Director of that Institute. The details of rice germplasm conserved in the Rice Research Stations of Tamil Nadu are given in Table 2.3.

To revive the cultivation of traditional varieties of rice, many private organizations and Non Government organizations started collecting landraces of rice from different places of Tamil Nadu, where the landraces are still in cultivation and also from farmers, local seed collectors and maintainers, growers who still keep seeds of indigenous rices etc; They display the seed materials in the local exhibitions organized during festival times, other special occasions and farmers days and explain the merits of traditional varieties and motivate many farmers to resort cultivation of traditional rices by sparing the seeds to the needy farmers for free of cost or at less cost.

Table 2.3: Rice Germplasm Conserved in Tamil Nadu

Name of the Research Station	No of Rice Germplasm
Paddy Research Station, Coimbatore	2508
Tamil Nadu Rice Resaerch Institute, Aduthurai	1450
Rice Research Station, Ambasamudram	150
Agri. Research Station, Thirupathisaram	150
Rice Research Station, Tirurkuppam	243
Coastal Salinity Research Centre, Ramanathapuram	26
Agricultural Research Station, Paramakudi	10

Subramanian *et al.*, 2010.

Utilization of Rice Landraces

The rice research which was started as early as 1902 in India resulted in the release of many high yielding rice varieties/hybrids to enhance the rice production and productivity in this country to meet the food demand of current and future domestic population and also to export to the needy countries. The rice breeding was started initially by domestication of wild rices followed by selection from landraces resulting in traditional varieties, subsequently the modern varieties. The improvement of all desirable characters through pureline selection was not been possible, hence hybridization technique adopted, which helped the breeders to combine and pyramid many needy and desirable traits scattered in many indigenous varieties in the deficit rice genotypes.

This could be possible because the vast variations found in the rice biodiversity for many desirable characteristic features. The higher yield, resistance to pests and diseases, tolerance to abiotic stresses and quality grain traits with high nutritive values available in the landraces of rice have been utilized to improved rice genotypes by adopting improved breeding technologies, This attempts have resulted in the release of many new improved rice varieties with high values for yield, nutrition, medicinal properties resistance to many biotic and abiotic stresses.

The first rice improvement work in the state was initiated as early as 1902 through Pureline selection from landraces/traditional varieties and released for cultivation. The particulars of Purelines selected from indigenous types in Tamil Nadu are furnished in the Table 2.4, by hybridization in Table 2.5 and a mutant is also obtained from a land race (Table 2.6).

Table 2.4: Pureline Rice Selections from Landraces/Traditional Varieties

Name of Rice Variety	Source of Selection	Name of Rice Variety	Source of Selection
ADT 1	Redsirumani	ADT 2	White sirumani
ADT 3	Kuruvai	ADT 4	Kuruvai
ADT 5	Nelloresamba	ADT 6	Redottaden
ADT 7	Whiteottaden	ADT 8	Early white sirumani
ADT 9	Poonkar	ADT 10	Korangusamba
ADT 11	Nelloresamba	ADT 13	Sannasamba
ADT 14	Vellaikar	ADT 15	Senkuruvai
Paddy Breeding Station, Coimbatore			
GEB 24 (Mutant)	Kichilisamba	CO 1	Periakichikli
CO 2	Poombalai	CO 3	Vellaisamba
CO 4	Anaikomban	CO 5	Chinnasamba
CO 6	Sadaisamba	CO 7	Sadaisamba
CO 8	Anaikomban	CO 9	Karsambared
CO 10	Gobikar	CO 11	Ayyansamba
CO 12	Thillainayagam	CO 13	Arupathamkodai
CO 16	Bonthamolagolukulu	CO 17	Chinnavadansamba
CO 18	Vellaikar	CO 19	Chingleputsirumani
CO 20	Tellasannavadlu	CO 21	Arupathamsamba
CO 22	Manavari	CO 23	Rangoonsamba
CO27	Pudupattisamba	CO 28	Bangarutheegalu
CO 32	Thiruchengodsamba		
Rice Research Station, Ambasamudram			
ASD 1	Karsambared	ASD 2	Karsambared
ASD 3	Veedhivadangan	ASD 4	Kuruvaikalayan
ASD 5	Karthigaisamba	ASD 6	Anaikomban
ASD 7	Karsambared	ASD 8	Thuyamalli
ASD 9	Avasarasamba	ASD 10	Kolavalai
ASD 13	Arikiruvai		
Rice Research Station, Tirur			
TKM 1	Pishini	TKM 2	Sembalai
TKM 3	Sornavari	TKM 4	Yerrasannavadlu
TKM 5	Manakattai		
Rice Research Station, Thalaignayar			
TNR 1	Thiruthuraipoondikar	TNR 2	Kattuvanam
Research Station, Palur			
PLR 1	Garudansamba	PLR 2	Chitraikali

Table 2.5: Landraces in Rice Improvement by Hybridization

Cross	New Rice Variety
ADT8	Molakolukuku/ADT2
TPS1	IR8/Kattisamba
MDU1	IR8/Chitralkar

Table 2.6: Natural Mutant from Landrace

Cross	New Rice Variety
Konamani	GEB24

Subramanian and Wifred Manual 1998.

References

Ahuja, S.C., Panwar, D, V., UmaAhuja, S., and Guptaka, 1995 Basmati Rice, the scented Pearl CCS Haryana Agricultural University, Hisar, Haryana.

Anonymous., 2009-2010. Crop and Season Reports Government of Tamil Nadu.

Anurudh K.Singh, 2013. Probable Agricultural Biodiversity Heritage sites in India XVII The Cauvery region Asian Agri-History; 17(4): 353-376.

Chellam, V., 2010.SangaIllakiathilValanmai. In: Nanjilla Velanmai, Special issue, Kumari Mavatta Nanchilla Velanmai Sangam 11th Conference pp. 42-43.

DabelDeb., 1995.Sustainable Agriculture and Folk varieties of Ecological, Economical and Cultural aspects. Mimco WWF Indian Eastern Region, Calcutta.

DabelDeb, 2000.Folk Rice varieties of WestBengal, Agronomic and morphological characteristics. Research Foundation for Science Technology; ed. FyRFSTE/Vihiri NewDelhi.

DeepakkumarRijal, 2010. Role of food tradition in conserving crop landraces on farm.*The Journal of Agriculture and Environment* ; 11: 107-119.

Frankel, O.H., and Hawkers, J.G., 1975. Genetic Research for Today and Tomorrow. International Programme Cambridge eds;Cambridge University Press, Cambridge, London.

FAO., 1999b. Women Users, Preservers and Managers of Biodiversity, Rome.

Harlen, J.R., 1975. Crops and man. American Society of Agronomy and Crop science Society of America Madison Wisconsin Science; 174: 468-474.

Khatana, V., Roy, J.K., and Pradhan.J., 2004.Collection of traditional varieties Western Odisha Lively world Project.Prepared with the assistance of DFID and Government of Odisha, Government of Odisha Department of International Development Western Odisha Rural Curlyhood Project.2004.

Madhanmohan, M., Balakrishnan, A., and Ranganayagi, P.R., 2013.A high yielding Seeragasamba rice culture VG 09006 and its medical properties. *Electronic Journal of Plant Breeding*; 42(2): 1148-1154.

Myers, N., 1994.Protected areas protected from a greater What? *Biology and Conservation* ;3: 411.

Personal communication from Deptt;of Agriculture and Research Stations of TNAU, 2004.

Personal Communication from ParialaThirumurugan., 2014. Chengalpattu, Tamil Nadu.

Rajukannu, A., Sathya, A., and OswaldQuintal, 2009.The diversity of traditional rice varieties in India: A Focus on Tamil Nadu PANAP Rice sheets, Publsiherd by Pesticide Action Nectuna Asia and Pacific, Malaysia.

Ramasamy, A., 1972.(Ed) Tamil Nadu District Gazettiers. Ramanathapuram Gazetters Madras, India.

Ravisankar, T., and Selvam, V., 1996.Contributions of tribal communities in the conservation of traditional cultivars. In: Using Diversity: Enhancing and Maintaining Genetic Resources Onfarm. International Develoment Research Centre, New Delhi, India.

Richharia, R.H., and Govindasamy, S., 1990.Rices in India 2nd Edition Academy of Development Science, Karjat Taluk Kashele, Maharastra.

Saravanan, E.P., 2012. Old landraces of rice and their medicinal properties. *Poorveegam* (a Tamil Journal); 8(3): 27-31

Sathya, A, 2013.Are the Indian rice landraces aheritage of biodiversity to reminiscence their past or reinventor future? *Asian Agri-History*, 17 (3): 221-232

Sathya, A, 2014.The art of naming traditional rice varieties and landraces by ancient Tamils. *Asian Agri-History*, 18 (1): 5-21

Subajkumarchoudhari., Genetic erosion of Agricultural biodiversity in India and Intellectual Property Rights Interplay and Key issues. email.subujceyahoo. co.uk.

Subramanian, M., and Wilfredmanual, W., 1998. Varietal Description of Rice strains released in Tamil Nadu and Pondicherry, Tamil Nadu Rice Research Institute, Aduthurai, Tamil Nadu, India.

Subramanian, M., Thiagarajan, K., Kalaiyarasi, R., andSundar, S., 2004.Genetic improve ment of Rice varieties in Tamil Nadu. In: Genetic Resources of Rice in India. ed. S.D. Sharma and U.Prasadrao. Today and Tomorrow Printers and Publishers and Publication, New Delhi pp. 937-1009.

Subramanian, M., Robin, S., and Thiagarajan, K., 2010.Ricegermplasm: Conservation and utilization in Tamil Nadu. In: Genetic Resources of Rice in India ed. S.D.Sharma, Today and Tomorrow Printers and Publishers, New Delhi pp. 391-429.

Subramanian, M., Robin., S.MurugesaBoobathy, P., and Natarajan., N, 2011.Role of indigenous rice varieties in rice improvement. In: Nel (Tamil book), ed. Subramanian.M., Robin.S., Murugesa Booba thy.P., and Natarajan.N.Sri Sakthi Promotional Litho Processes, Coimbatore.

Vaughan, 1994.The wild relatives of Rice: A Genetic Handbook. International Rice Research Institute, Manila.

Vetelainen, M., V.Nagari, and Maxted.N., 2009.European landraces onfarm Conservation and Managemant and Use; Biodiversity Technical Bulletin, N0 15 Published by Biodiversity International.

Vijayalaksmi, K., and A.Nambi., 1997.Toward setting up a community seedbank: Experience from Chengam farm Tamil Nadu. In: Sperling, L. and M.Loevinsha (eds.) Using diversity. Enhancing and mainataining Genetic Resources On Farm.IDRC (International Development Research Centre)NewDelhi India pp. 281-288.

Zeven, A.C., 1998.Landraces.A review of definitions and classifications. *Euphytica*, 104: 127-139.

Chapter 3

Rice Production in Tamil Nadu: Challenges and Opportunities

R. Rajendran

Director Incharge,
Tamil Nadu Rice Research Institute,
Aduthurai – 612 101, Tamil Nadu
e-mail: rajendrankmu@yahoo.co.in

Tamil Nadu is one of the major agrarian states in India where agriculture plays a significant role for rural employment, income and livelihood of people. The total domestic agricultural production of Tamil Nadu state is 9.16 percent, mainly contributed by agriculture which in turn provides 60 per cent of job opportunity. Rice is the staple food crop, cultivated in 33 per cent of total cultivable area. Rice is cultivated in an area of about 21 lakh hectares with a production of 65 lakh tonnes and productivity of 3.1 tonnes/ha. Rice is cultivated throughout the year in three major seasons *viz., Kar/Kuruvai/Sornavari* (16.6 per cent) season (April to July), *samba/ thaladi/pishanam* (August to October) season (76.5 per cent) and navarai/summer – 6.8 per cent (January to May). Rice is cultivated in all the districts of Tamil Nadu except Chennai. In particular, Cauvery delta zone comprising of three districts *viz.,* Thiruvarur (8.9 per cent), Thanjavur (8.8 per cent), and Nagappatinam (8.7 per cent); Viluppuram (7.9 per cent) and Ramanathapuram (6.9 per cent) are other important districts cultivating rice in more area.

Tamil Nadu is one of the leading rice growing states in India, especially in the Cauvery delta region where rice is cultivated in about 35 per cent of Tamil Nadu rice area contributing major share in rice production and hence proudly called as "Granary of Tamil Nadu" state. However, in recent years rice cultivation is facing several constraints and challenges posing great concern on sustaining rice production. The last ten years of rice production scenario revealed the need for sustainable rice production to meet the demand of growing population. But burgeoning population beyond food production levels is sending signals to take appropriate measures for enhancing food grain production, more particularly rice. Dwindling rice area due to urbanization, water and labour scarcity, climate change, declining soil fertility and escalated input cost discourages rice farmers of the state and whole of Asia. The problems associated with rice production are diverse and location-specific *viz.*, declining farm resources (falling of groundwater, encroachment and silting of reservoirs), deteriorating soil quality (chemical pollution, salinity build up) and increased competition from other sectors such as urban and industrial users. Since demand for rice is still rising because of burgeoning population in Asia, there is an urgent need to "grow more rice with less water" (Guerra *et al.*, 1998). The most disheartening factor is the lack of interest among the younger generation on agriculture and more specifically on rice farming.

Current Challenges

Industrial growth, diversified demand of growing population resulted in reducing farm area and increase in the non-agricultural land use. In Tamil Nadu, gross cropped area during 2000 – 2001 was 63.38 lakh ha, which was reduced to 55.72 lakh ha in 2009 -10. The net sown area of 61.3 lakh ha in 1970, reduced to 48.92 lakh ha during 2009 – 2010. This situation is forcing us to produce more food on less land.

Out of the total water used for agriculture, about 70 per cent is used only for rice cultivation. On the other hand, industrial, municipal, and livestock sectors increased their share of water use from 13 per cent in 1968 to 37 per cent in 2000. On an average rice requires 3000 – 5000 lit of water to produce a kilogram of grain. Reduction of area is much faster in rice than other crops due to its high requirement of water. Climate change, increasing population and industrialization are the major causes for shortage of water to agriculture and more so for rice cultivation. In Cauvery delta region, rice productivity and production dwindled drastically during 2012 - 2013 due to failure of North East monsoon rain and non-availability of Cauvery water. The fast declining ground water and deteriorating water quality also threatens rice cultivation in few parts of Cauvery delta and Tamil Nadu state as a whole. In rural areas, groundwater table has been steadily going down by about 0.8 m between 1996 and 2002 in the upstream part and by about 2.8 m in the downstream part. The amount of water available for irrigation, however, is also becoming increasingly scarce due to competition from non-agricultural water users. It is essential to "produce more rice with less water" for sustainable food production. In Asia, rice is largely established by transplanting (Pandey and Velasco, 1999) which consumes more water starting from nursery rising to transplanting and harvest. Water-saving irrigation technologies for rice are seen as a key component in any strategy to deal with water scarcity (Li and Barker, 2004). The strategic research on water saving rice crop establishment is the need of the hour.

Of the total cost of rice cultivation, about 45 - 50 per cent of goes to labour cost due to escalating wages of farm labour and scarcity due to labour migration from villages to cities for non – agriculture works. During the peak season, farmers are in need of labour force at the specific period of time to complete their farm operations *viz.*, transplanting, weeding and harvest.

Declining Soil Fertility

Rice cultivation is practiced by Cauvery delta farmers from time immemorial. Soil fertility and productivity has declined due to mono cropping and less preference for growing other crops, reduced application of organic manures and imbalanced and indiscriminate use of inorganic fertilizers. Decline in soil organic carbon content leads to poor nutrient use efficiency of inorganic fertilisers. Most rice soils in Tamil Nadu are deficient in organic matter and micronutrients especially zinc content. Tamil Nadu soils are deficient in micronutrients *viz.*, zinc (52 per cent), iron (28 per cent), copper (7 per cent) and manganese (6 per cent). Sodic soils and coarse textured calcareous soils with low organic matter content suffer from Fe deficiency, besides Zn and Cu (Savithri *et al.*, 1999). Low organic matter content in soil reduces the biological activity, water holding capacity and nutritional availability leading to poor rice productivity. In order to improve the soil health and soil fertility, application of bio-fertilizer, cultivation of green manure crops, use of composted farm wastes need to be given top priority.

Problem Soils

In Tamil Nadu, coastal tract of Thanjavur district in Cauvery delta has an area of about 0.18 million ha along the sea shore of about 280 km which is frequently intruded by marine water causing salt accumulation (Saravanan *et al.*, 1991). Coastal soils pose severe physical constraints in association with their inherent site characteristics such as dry, humid climate, texture and single grain soil structure. Sodic soils have poor physical and chemical property, impeded water infiltration, water availability and ultimately plant growth. In many coastal areas uncontrolled mining of groundwater had resulted in intrusion of seawater and development of high salinity (Velayutham *et al.*, 1999).

Saline soils have low organic matter and the efficiency of N fertilizer usage by crops is poor in saline soil. Availability of P increased up to a moderate level of salinity, but later it decreases. The coastal soil shows K deficiency due to antagonistic effect of Na and K absorption and or disturbed Na/K ratio.

Improper water management, increased usage of ground water and intrusion of sea water are causing more salinity and alkalinity in soil which affects the soil physical and chemical properties resulting in reduced soil infiltration, water use efficiency leading to poor rice yields.

Climate Change

Global warming as a result of climate change is the foremost challenge for agriculture including rice cultivation. Climate change affects the rice productivity through various forms *viz.*, cyclone, flood, drought and temperature fluctuations. Furthermore, increasing night time temperature affects rice growth and yield due to high demand of energy for respiration.

Biotic Stresses

In rice cultivation, 20 – 30 per cent of yield loss is occurring due to weeds. Labour shortage and escalating wages are the major factors that mostly affect rice farmers especially when rice is direct seeded. Improper land levelling and water shortage increases the weed problem and reduces rice yields.

Integrated pest management practices have been developed for major pests in rice. In the changing climatic scenario, few minor pest and diseases cause severe damage to rice crop in recent years especially infestation by mites and thrips from the nursery to main field and growing incidences of false smut irrespective of the rice varieties grown. Recently, severe mealy bug infestation noticed in dry direct seeded rice. Increasing infestation of false smut, bacterial blight and rat damage are the major concern for our scientists, extension workers and farmers. In addition to enhanced cost on plant protection, increased use of pesticide causes environmental pollution. Intensive cultivation and agro climatic factors aggravate the pest and disease problems. Lack of awareness and knowledge on rice and pest incidence still persist among the rice growers of Tamil Nadu which needs to be addressed through mass campaigning and trainings.

In rice, blast, bacterial blight, sheath rot, sheath blight and grain discolouration affect rice crop often causing economic loss to the farmers. The marketable quality of the grains become poor and fetches low price. A number of weak parasites/ saprophytes which infect rice seeds at pre and post harvest stages cause grain discolouration. The types and magnitude of discolouration vary with the place and environmental conditions of the organisms involved. Weather elements such as continuous rain, humidity and temperature conducive to fungal and bacterial growth increase the incidence of grain discolouration. Rice growers are facing so many challenges that reduce the interest among the farming community. Therefore, there is an urgent need to disseminate farmer friendly technologies to solve the problems faced by the rice growers in Tamil Nadu.

Opportunities and Strategies

Enhancing Soil Fertility

Improving the organic carbon content of soil, is the foremost thing to be achieved thro' adoption of rational agronomic practices to increase the soil fertility and productivity. Addition of available organic manures, green manure and green leaf manure are to be given top priority for enhancing the rice productivity of the state. Incorporation of suitable green manure suiting to the existing rice eco –system should be promoted for enhancing soil fertility and productivity. *Sesbania aculeate* (daincha/thakkaipoondu) and *Sesbania rostrata* (manila agathi) are suitable green manures for wet clay soil conditions. *Crotolaria juncea* (sun hemp/sanappu) is the suitable green manure for light sandy loam soils having adequate drainage. Emphasis should be given for adding available crop residues in the rice based cropping system. The chaffed paddy straw spread by the combine harvester can be incorporated instead of burning the paddy straw in the field. Tamil Nadu Agricultural University has released microbial consortia in the name of bio – mineraliser for early decomposition of crop residues in the field. Agronomic practices encouraging crop residue incorporation for example, inter-cultivation with weeders as recommended for SRI should be promoted as it will increase soil fertility in addition to weed burrial.

Judicious Water and Nutrient Management

Water is the single most important and precious input for sustainable rice production, especially in the traditional rice growing areas of Tamil Nadu. Reduced investments in irrigation infrastructure, increased competition for water and large withdrawals from underground water source now threaten the rice production in many parts of the state and Asia as a whole. Producing more rice with less water is therefore a formidable challenge for the food and livelihood security of the region. Water management is another area where much attention is needed on the extension side. Tapping the underground water for agriculture is gaining momentum in the delta zone to supplement the water needs. The uncertainty and limited availability of water in the canals created awareness on economic use of irrigation water. The delta farmers already started thinking of alternatives in *lieu* of traditional crops. Water use efficiency is to be taught through larger demonstrations and preferably taking village as a whole.

The daily consumptive use of water for rice varies from 6-10 mm and total water ranges requirement ranges from 1100 to 1250 mm depending upon the agro climatic situation, duration of variety and soil characteristics. Good land levelling minimizes the water requirement. Continuous submergence not only consumes more water but also reduce rice yields as experienced thro' innovative crop establishment approaches *viz.*, SRI, dry direct seeded rice and wet seeded rice. Intermittent irrigation at regular intervals is the one solution recommended especially in delta regions. Irrigation interval is adjusted in such a way that the crop will not suffer due to water deficit at any growth period. Farmers are experiencing intermittent drought for the past few years farmers and understood the benefits of Alternate Wetting and Drying (AWD) method of irrigation. Direct dry seeding is being recommended wherever possible to save 25-30 per cent of water as this approach avoids nursery preparation, puddling of soil and transplanting which consumes considerable quantity of water. In ideal dry season with assured water availability, SRI method of cultivation with AWD system can be recommended which gives 15 – 20 per cent higher yield with a saving of about 25 per cent water.

Balanced nutrition to rice is the key for enhancing rice productivity. TNAU has developed technologies for need based fertilizer management in rice. Rice nutrition is done rather recommended based on soil analysis, DISSIFER software developed by TNAU and many other ways. Site Specific Nutrient Management (SSNM) was released as a technology in 1998 for rice nutrient management in Cauvery delta. SSNM approach supplies the plant nutrient as required by rice crop which takes in to account of the indigenous nutrient supplying capacity of the soil. A Nutrient Manager for Rice (NMR) software was also developed by IRRI, Philippines in association with TNAU based on the decade of SSNM research conducted in Cauvery delta region. NMR is under final stage of refinement and evaluation in Cauvery delta for need based NPK fertilizer application at individual field level. Bio-composts possess macro and micronutrients, besides having the ameliorating effects to improve the soil properties of salt affected coastal soils and it can be converted as an ideal manure with higher content of macro and micronutrients. Application of enriched pressmud @ 20 t ha^{-1} recorded changes in physico-chemical properties in alfisol followed by a reduction in pH and EC and an increase in the organic carbon content of soil. Application of enriched pressmud compost increased the exchangeable Ca and Mg and micronutrient status of soil (Dhanushkodi and Subrahmaniyan, 2012).

Mitigating Climate Change Effects

With the likely growth of world's population towards 10 billion by 2050, the demand for rice will grow faster than for any other crops. There are already many challenges exist to achieving higher productivity of rice. Now new challenges add to agriculture which includes changing climate and its consequences. At the end of the twenty-first century, the increases in surface air temperature will probably be around 1.4–5.8 °C, relative to the temperatures of 1980–1999. Most of the rice is currently cultivated in regions where temperatures are above optimal for crop growth (28/22°C). Any further increase in mean temperature during sensitive stages may reduce rice yields drastically.

In tropical environments, high temperature is one of the major environmental stresses limiting rice productivity, with relatively higher temperatures causing reductions in grain weight and quality. Developing high-temperature tolerant rice cultivars has become a proposed alternative, but requires a thorough understanding of genetics, biochemical, and physiological processes for identifying and selecting traits, and enhancing tolerance mechanisms in rice cultivars. Most agronomic interventions for the management of high-temperature stress aim at early sowing of rice cultivars or selection of early maturing cultivars to avoid high temperatures during grain filling. But these measures may not be adequate as high-temperature stress events are becoming more frequent and severe in the future climate. Suitable varieties should be selected based on excess or scanty rainfall and altered temperature. Drought tolerant rice varieties *viz.*, Anna (R) 4 and PMK 3 are popularized in drought prone areas of Tamil Nadu. Besides drought and high temperature, excess rain, flash flood and submergence also cause considerable rice yield loss now a day. IRRI has developed the flood tolerant Swarna sub-1 rice variety suitable for flood prone areas. The performance of this variety is being assessed in low lying areas of Thiruvarur and Nagapattinam districts in Cauvery delta.

TNAU in association with Norway is operating a project called CLIMARICE for promoting farmer friendly technologies to mitigate the ill effects of climate change in Tamil Nadu. We need to create awareness among the farmers regarding climate change and take strategic decisions for escaping the ill effects of climate change. Further, daily farming activities and decisions have to be made based on instant weather as much as possible. Tamil Nadu Rice Research Institute, Aduthurai in association with Indian Meteorological Department is periodically giving the medium range weather forecast with necessary farm advisory to seven districts of Cauvery delta. Therefore, such interventions could likely to increase the yield and reduce the negative impact of climate change on rice.

Managing Weeds

Weeds cause problems in all types of field and pollute and deplete water bodies besides hindering with rice production activities. Weeds intensify pest and disease problem by serving as alternate host, reduce the land value and reduce rice yields and quality. Integrated weed management approach is the solution for sustaining the crop free from weed problem. Herbicide application becomes inevitable in selected rice ecosystems especially under direct seeded conditions. Pre-emergence herbicide application is more preferable as weeds are controlled at the time of emergence itself. Post emergence application can follow only on need basis depending upon the type of weed flora. Butachlor, anilophos, pretilachlor plus safener, pendimethalin and pyrozo sulphuron ethyl are some of pre emergence herbicides recommended for rice. In case of dry direct seeded rice, pendimethalin as pre-emergence application effectively control the grass and broad leaved weeds. Care should be taken when we select herbicide for post emergence application. 2.4-D Na salt is the common post emergence herbicide used to control broad leaf weeds in rice. In dry direct seeded rice, bispyribac sodium is recommended as post emergence spray to mange grass weeds and sedges. Farmers need to be enriched with knowledge on weed management strategies. Selection of suitable herbicide, dose, time of application, stage of crop, soil moisture condition in case of pre-emergence application are some of the important factors to be kept in mind for effective and safe management of weeds.

Mechanized Rice Farming

Few years back, there was resistance from agricultural labourers for the introduction of machineries in the field. The labour scarcity and escalated wages forced the farmers to go in for mechanized rice cultivation. The efficiency of farm labourers also reduced since 1990s and the usage of machineries increased from 7.8 to 14.7 hours per hectare.

Mechanization in rice cultivation is the need of the hour especially during the peak transplanting and harvesting periods. Transplantation is slowly picking up in the Cauvery delta and Tamil Nadu as a whole. Almost 70-80 per cent of rice crop is harvested by combine due to high labour wages and labour scarcity. But still, in dry direct seeded rice, use of tractor drawn seed drill has started picking up since 2011 only. In most of the direct seeded rice areas, farmers sow the seeds by broadcasting only. In tractor drawn seed drill sowing, optimal plant population is maintained which facilitate for inter cultivation using cono weeder. Under the broadcast method, farmers use excess seed rate and rigorous thinning becomes essential which increase the labour cost apart from loss of considerable quantity of precious seed material. Because of the introduction of SRI in Tamil Nadu, there was a tremendous change in the mind set of farmers in reducing the seed rate. A farmer in Nagapattinam district is challenging to grow rice with the lowest seed rate of 625 g/ha where as the recommended seed rate for SRI method of cultivation is 5 -7 kg/ha. The seed use also came down for the machine transplanted rice. The seed rate for mat nursery came down to 20 kg/ha from 37.5 kg/ha, half the seed rate reduced because of sound rice mat nursery technologies followed by the entrepreneurs and farmers. TNAU released the modified rice mat nursery technology in 2006 suitable for planting single seedling under the SRI method (Rajendran *et al.*, 2005). Use of cono/rotary weeder need to be encouraged as it incorporates the weed in to the soil besides creating soil aeration. Stirring of soil in line planted crops especially during the cold weather period aerate the root zone area and encourage rice crop to produce more number of tillers. Large scale adoption of line planting and mechanized inter cultivation could make sizable increase in grain yield of rice. In line sown dry seeded rice also, mechanized inter cultivation could play a significant role in enhancing tiller production and grain yield. Use of disc plough/chisel plough once in 3-4 years would break the hard pan formed in heavy clay type of soil because of the hard sub soil formed by intensive puddling being done for years together. Land levelling is one of the key management issues which need serious attention. As far as possible, laser land levelling can also be encouraged to conserve soil, nutrients, water and increase rice productivity. Application of engineering principles for reducing energy requirement in the form of human, animal, mechanical and electrical power is necessary to reduce cost of rice production and agriculture in general. Efficient tools and implements suitable for medium to small size land holdings are to be designed to reduce drudgery of labourers and animals and to reduce time and cost. Appropriate post-harvest practices for cleaning, grading, drying, processing and storage are needed to improve the quality of grain and by-products.

Managing Problem Soils

Selection of saline/sodic tolerant rice variety is the pre requisite for getting higher rice productivity. The recently released rice variety TRY 3 is found to be tolerant to saline and sodic soil conditions besides its suitability for idly making. Varieties *viz.*, TRY 2, CO43 are also suitable for saline soil conditions. Reclamation of sodic soils with gypsum application and green manure incorporation helps to enhance rice productivity. Application of 25 per cent extra nitrogen and zinc is recommended for problem soils to get sustainable rice yields.

What Should be done to Enhance Rice Productivity?

All the stake holders involved in rice industry, scientists of Tamil Nadu Agricultural University, state extension officers and farmers have to join together and work to overcome the challenges and enhance rice production and profit taking into account of the ultimate consumer preference.

Researchers

Universities not only release newer technologies for rice but also many new varieties of other crops to help the country attain self-sufficiency in food production. However, researchers should focus their attention on need based farmers' issues and identify strategic solutions for friendly use by the farming community. Researchers are responsible to develop eco-friendly, high yielding with low input responsive rice varieties, and hybrids considering the eco- system and environment. The locally adapted varieties need to be introgressed with genes resistant to pest and diseases isolated from traditional land races for sustaining the rice productivity. Also scientist should concentrate on developing medicinal rice varieties and varieties suitable for preparation of value added products like popped rice, rice flakes etc.

Extension Workers

Extension officers are the bridge connecting researchers and farmers. The feed back received from extension officers would strengthen the research activities of the scientists for developing farmer friendly varieties and technologies to increase rice production of any region. Adequate training should be given to extension officials on selected technologies which in turn enrich the knowledge of contact farmers and create a chain of knowledge sharing link.

Farmers

Although Universities release varieties and agro technologies for the benefit of farmers, it is up to the farmers to make use of all those findings with interest and involvement. Small farmers should join together in groups to enrich their knowledge and maximise their farm income. Excessive applications of fertilizer, over watering, indiscriminate use of plant protection chemicals should be avoided in future. Farmer groups could make more progress and achieve their goal on activities like community mat nursery, rat control, water management etc. Farmers could get bumper harvest and maximise their income and livelihood security through group approach only.

References

Dhanushkodi, V. and K. Subrahmaniyan. 2012. Soil Management to Increase Rice Yield in Salt affected Coastal Soil - A Review. *Int. J. Res. Chem. Environ.* 2(4): 1-5.

Guerra, L.C. S.I. Bhuiyan, T.P. Tuong and R. Barker. 1998. Producing more rice with less water from irrigated systems. SWIM paper 5, IWMI/IRRI, Colombo, Sri Lanka. 24 pp.

Li, Y. and R. Barker. 2004. Increasing water productivity for paddy irrigation in China. *Paddy Water Environ.*, 2: 187 - 193.

Pandey, S. and L. Velasco. 1999. Economics of direct seeding in Asia: patterns of adoption and research priorities. *Mini Rev. Int. Rice Res. Notes* 22: 6 – 11.

Rajendran, R., Ravi, V., Valliappan, K., Nadanasabapathy, T., Jayaraj, T., Ramanathan, S. and V. Balasubramanian, 2005. Early production of robust seedlings through modified mat nursery for enhancing rice (*Oryza sativa*) productivity and profit. *Indian J. Agron.* 50(2) : 132-136.

Saravanan, A., P. Malarvizhi and Ramanathan. 1991. Use of organic amendments for improving rice yields in seawater intruded soils of Cauvery delta. *J. Indian Soc. Coastal agric. Res.*, 9(1 and 2): 203-207.

Savithri, P., R. Perumal and R. Nagarajan. 1999. Soil and crop management technologies for enhancing rice production under micronutrient constraints. *Nutr. Cycl. Agroecosys, 53:* 83 - 92.

Velayutham, M., D. Sarkar, A. Natarajan, P. Krishnan, C.R. Shiva Prasad, O. Challa, C.S. Harindranath, R.L. Shyampura, J.P. Sharma and T. Bhattacharyya. 1999. Soil resources and their potentialities in coastal area of India. *J. Indian Soc. Coastal agric. Res.*, 17(1 and 2), 29-47.

Chapter 4

Rice: Genesis and Improvement

R. Saraswathi, P. Shanthi, R. Suresh and R. Rajendran

Tamil Nadu Rice Research Institute,
Aduthurai – 612 101, Tamil Nadu
e-mail: arsvpm@tnau.ac.in

Rice has fed more people over a longer period of time than any other food crop. Far back as 2500 BC, rice has been documented in the history books as a source of food and for tradition as well. Rice crop originated in China and was spread to surrounding countries such as Sri Lanka and India. It was then passed onto Greece and West Asia in 300 BC by the armies of Alexander the Great (www.cgiar.org). In 800 AD, people in East Africa traded with people from India and Indonesia and were introduced to rice.

Genetic evidence had shown that rice originated from a single domestication 8200 to 13500 years ago (Molina *et al.*, 2011) in the Pearl River valley of China. Previously, archaeological evidence had suggested that rice was domesticated in the Yangtze river valley region in China (Chang, 1976).

From East Asia, rice was spread to Southeast and South Asia; rice was then introduced to Europe through western Asia and to the America through European colonization (Plate 4.1). Rice could be taken to many parts of the world due to its versatility. It can be grown in a multitude of conditions, starting from desert lands of Saudi Arabia to wet land deltas of South East Asia. It is grown all over the world except Antarctica. It is grown in four different ecosystems *viz.*, irrigated, rainfed lowland, rainfed upland and flood prone areas (Plate 4.2).

People have used rice to make snacks, desserts, main courses, alcoholic beverages and special foods for religious ceremonies. Rice is rich in carbohydrates and provides instant energy. For millions of people, rice is three fourth of their diet *i.e.* 35-80 per cent of total calories intake (Yan *et al.*, 2001). In husked rice, protein content ranges between 7.0 and 9.0 per cent. The brown rice (unpolished rice) is rich in some vitamins, especially B_1 or thiamine (0.34 mg), B_2 or riboflavin (0.05 mg),

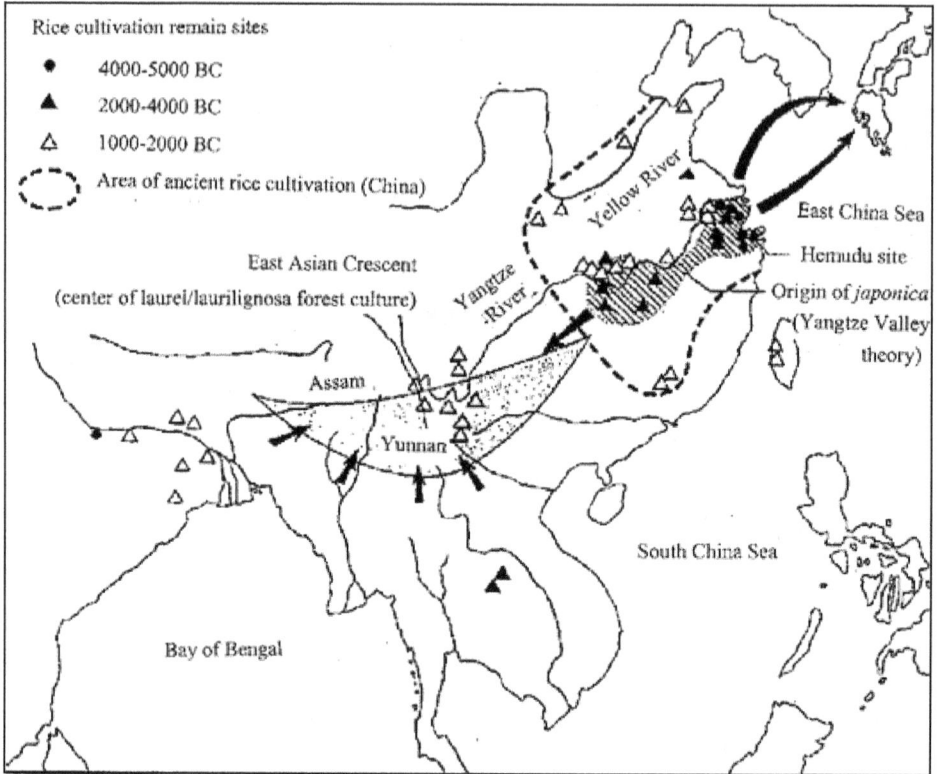

Plate 2.1: Domestication of Rice in Yangtze River Valley.

Plate 2.2: Up Land Rice and Wetland Rice.

niacin or nicotinic acid (4.7 mg). Besides, some of the rice varieties like Kavuni and Illuppaipoo Samba (Tamil Nadu), Kanthi Banko (Chattisgarh), Meher, Saraiphul and Danwar (Odisha), Atikaya and Kari Bhatta (Karnataka) Navara, Kunjinellu, Erumakkari and Karuthachembavu (Kerala) have medicinal and curative properties.

Early Spread of Rice

In early periods somewhere in Asian arc, rice was grown in forest clearing under a system of shifting cultivation. Later, this crop was grown by direct seeding and without standing water. A similar but independent pattern of incorporation of wild races into an agricultural system might have taken place in one or more locations of Africa, approximately at the same time. Later in China, the process of puddling the soil and transplanting seedlings were likely refined. Transplanting provides the farmer with the ability to better accommodate the rice crop to a finite and fickle water supply by shortening the field duration (since seedlings are grown separately in the nursery) and adjusting the planting calendar. With the development of puddling and transplanting, rice became truly domesticated.

In China, the history of rice in river valley and low lying areas is longer than the history of its dry land crops. In Southern Asia, by contrast, rice was originally produced under dry land conditions in the uplands, and only recently did it come to occupy the vast river deltas (Plate 4.3).

Plate 4.3: Domestication of Rice.

Migrant people from South China or perhaps northern Vietnam carried the traditions of wetland rice cultivation to the Philippines during the second millennium B.C., and Deutero - Malays carried the practice to Indonesia about 1500 B.C. From China or Korea, the crop was introduced to Japan no later than 100 B.C. Movement to western India and south to Sri Lanka was also accomplished very early, *i.e.*, around 2500 B.C. for Mohenjo-Daro and in Sri Lanka, rice was a major crop as early as 1000 B.C. The crop might have been introduced to Greece and neighboring areas of the Mediterranean during the return expedition of Alexander the Great to India during 344-324 B.C (Plate 4.4).

The 16th century played an important role in limiting the adoption of rice as a major crop in the Mediterranean area. During the 16th and early 17th centuries, malaria was a major disease in southern Europe, and it was believed to be spread by the bad air from swampy areas. Major drainage projects were undertaken in

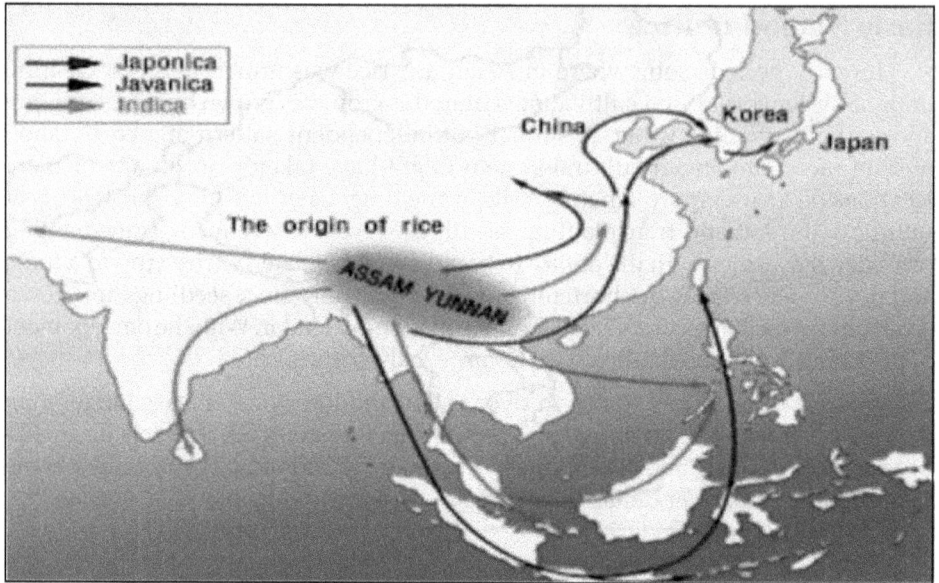

Plate 4.4: Early Spread of Rice.

southern Italy and wetland cultivation was discouraged in some regions. In fact, it was actually forbidden on the outskirts of a number of large towns. Such measures were a significant barrier to the diffusion of rice in Europe. As a result of Europe's great age of exploration, new lands to the west became available for exploitation. Rice cultivation was introduced to the New World by early European settlers. The Portuguese carried it to Brazil and the Spanish introduced its cultivation to several locations in Central and South America. Early in the 18th century, rice spread to Louisiana, but not until the 20th century it was produced in California's Sacramento Valley.

History of Rice in India

The *indica* rice was first domesticated in the area covering the foothills of the Eastern Himalayas (north-eastern India), stretching through Burma, Thailand, Laos, Vietnam and Southern China. The *japonica* variety was domesticated from wild rice in Southern China and then introduced to India before the time of the Greeks. Chinese records of rice cultivation dates back to 4000 years.

Rice is first mentioned in the Yajur Veda (1500-800 BC) and then frequently referred to in Sanskrit texts. Rice is often directly associated with prosperity and fertility in our country. It seems to have appeared around 1400 BC in Southern India after its domestication in the Northern plains. The word "rice" is derived from the old French word *ris*, which comes from Italian *riso* and in turn from the Latin *Oriza*.

Botany

Rice belonging to the family *Poaceae* and genus *Oryza*, has two cultivated species viz., *Oryza sativa* and *Oryza glaberrima*, of which the most widely cultivated one is

O. sativa consisting of three subspecies *viz., indica, japonica and javanica.* The genus *Oryza* consists of 23 species (Tateoka, 1962). The major rice varieties grown world wide today belong to *Oryza sativa indica* and *Oryza sativa japonica.* The two cultivated rice species, belong to a species group called *Oryza sativa* complex together with the five wild taxa, *viz., O. rufipogon* (sensu lato), *O.longistaminata* Chev. et Roehr., *O. barthii* A. Chev., *O. glumaepatula* Steud., and *O. meridionalis* Ng. The wild rice *O.rufipogon* and *O. barthii* are believed to be the immediate progenitors of the cultivated rice *O.sativa* and *O. glaberrima* respectively. Mutation is responsible for the key traits that made rice easier to cultivate. Most of the researchers have concluded that human directed hybridization played an important role in the very early history of domesticated rice (Harlan and Dewet, 1971) (Plate 4.5).

Plate 4.5: Wild Rice.

The common cultivated rice plant is an annual which usually grows to a height of half meter up to two meters. Some deep water rice varieties grow with the gradual rise of the flood water level. The root system comprises of secondary adventitious roots produced from the underground nodes of the culm. The shoot system is composed of culms, leaves and inflorescence (panicle). The hollow culm or stem is made up of a series of nodes and internodes. Each node bears a leaf and a bud. Buds near ground level grow into tillers. The primary tillers give rise to secondary tillers which give rise to tertiary tillers. The leaves of rice are sessile in nature. They are borne at an angle, on the culm in two ranks along the stem, one at each node. The leaf blade is attached to the node by the leaf sheath (Plate 4.6).

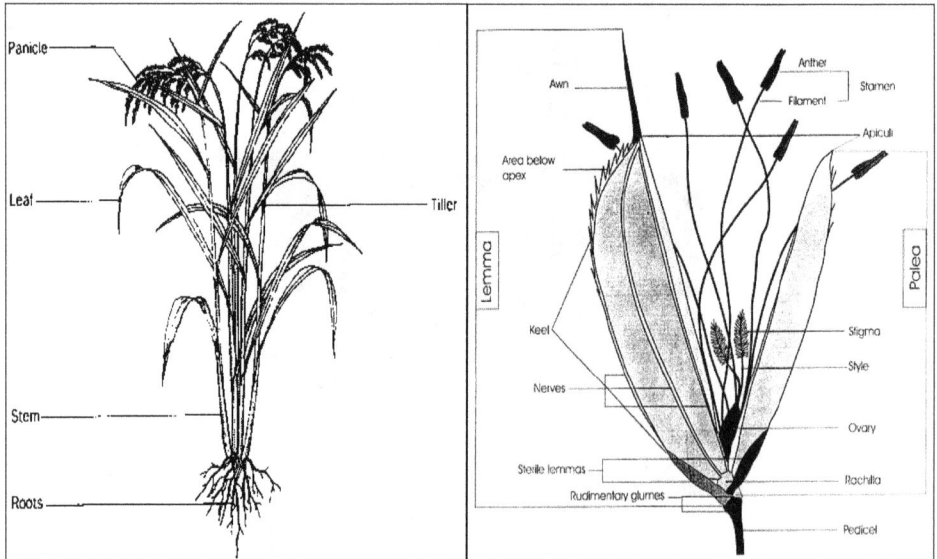

Plate 6: Morphology of Rice.

The inflorescence of rice is known as panicle, comprising the primary branch which is divided into secondary and sometimes tertiary branches. The individual spikelet consists of two outer glumes namely lemma and palea (hull) and the complete flower between them. The rice flower contains six functioning stamens (male organ) and a pistil (female organ). Rice is a self pollinated crop. When rice flower blooms, the lodicules at base of the flower become turgid and push the lemma and palea apart, thus allowing the stamens to emerge outside the open floret. Rupturing of the anthers then leads to the shedding of pollen grains and pollination.

Rice grain (caryopsis) develops after the fertilization is completed. The pericarp is the outermost layer which envelopes the caryopsis and is removed when rice is milled and polished. The embryo lies at the ventral side of the spikelet next to the lemma. Adjacent to the embryo is a dot like structure called hilum. The embryo contains the plumule and radicle. The plumule is enclosed by a sheath known as coleoptile and the radicle by the coleorhizae.

Rice Improvement in Early Periods

Early breeding research in tropical Asia began with the improvement of native varieties by 'pure line selection', the strategy for purification of highly heterogeneous farmer varieties into readily recognizable ones of uniform stature, maturity and grain type. Though the strategy had helped to restrict the number of varieties, it hardly helped to raise appreciably the genetic yield level. The rediscovery and elucidation of Gregor Johann Mendel's Laws of Inheritance, a hundred years back was the defining moment in the history of biology. It was the understanding of the laws of genetics that convinced crop breeders that hybridization was the most effective means to generate more variability, on which selection could be practiced in desired direction. Application of this knowledge in rice was however, restricted to improvement of simply inherited traits rather than the genetically complex yield. The pure line selection had helped to raise the yield level by 5-10 per cent over the landraces, the gain through hybridization had been another 10-15 per cent. Rice yields thus remained practically very low and stagnant.

In early 50's, FAO sponsored the *indica-japonica* hybridization project Asia – wide, to combine fertilizer responsiveness and non lodging plant habit of temperate *japonica* varieties with wide adaptability and acceptable grain quality of *indica* rice. In India, CRRI, Cuttack took up country wide screening of the segregating material as the primary hybridization centre. This 10 year long international initiative to raise the genetic yield level of *indica* rice proved practically a futile exercise. Except for four varieties of promise *viz.*, ADT 27 an early maturing type found suitable for *kuruvai* season in Tamil Nadu; two medium late maturing types Mahsuri and Malinja identified in Malaysia that became popular in India and Malaysia respectively and Circna in Australia, the intended objective could hardly be achieved.

Chinese breeders developed the first ever semi-dwarf high yielding variety Guang-Chang-ai in late 1950's using the dwarf gene from Ai-zi-Zhan (Hung, 2001). Parallel to this, Taiwanese developed Taichung (Native) using same dwarfing gene (SD1) from Dee-Geo-Wu-Gen (DGWG) in 1956. International Rice Research Institute at Manila, Philippines developed the IR 8 variety (yield potential up to 10 t/ha) by crossing DGWG with an Indonesian variety Peta which finally helped in breaking the yield barrier in rice using semi dwarfing gene (SD1). Similar concept was used in Korea and Japan. Tongil varieties developed in Korea largely with background of *japonica* genome, by crossing *japonica* and *indica* was short, non lodging with long and heavy panicles (Saiato *et al.*, 2007) (Plate 4.7).

Rice breeding in India was initiated by Dr.G.P.Hector in 1911 in Dacca (now in Bangladesh). In the following year *i.e.*, 1912, a Paddy Breeding Station was started at Coimbatore in the old Madras province to meet the requirement of South India, which is another major rice growing area of the country. Realizing the importance of rice to India's economy, the Imperial (Indian) Council of Agricultural Research, founded in 1929, sponsored rice breeding projects in all the major rice growing states and as a result the country had as many as 82 research stations exclusively devoted to rice research by 1950. Initially, most attention was paid for screening and selection of native rice so as to recommend few promising varieties of *indica* types.

Plate 4.7: Semi Dwarf Rice Varieties.

The Bengal Famine of 1943, the worst disaster of India during the twentieth century in which nearly four million people died of hunger in Bengal, prompted the Central government for production, procurement and supply of food. Therefore, it was decided to establish a Central Institute for Rice Research in the country and Indian Council of Agricultural Research (ICAR) established the Central Rice Research Institute at Cuttack, Odisha (now, Odisha) in 1946 for addressing the problems of national importance. Many Chinese, Japanese, Taiwanese and Russian type rice were introduced and tested. The Chinese types introduced, prior to 1947 and tested in Kashmir Valley were found fairly successful and the Japanese and Russian types were found unsuitable under Indian conditions due to poor yield, unacceptable grain quality and susceptible to blast (Das, 2012).

Before the establishment of the CRRI, the organized breeding research at central and state levels led to the release of more than 450 improved varieties, all of which, except 27 cross bred varieties were pure line selections from popular local varieties. This simple crop improvement strategy led to development of hundreds of genetically homogenous cultivars, which are now referred to as 'traditional cultivar germplasm'. While meeting the varietal needs for long, the 'cultivar germplasm' housing a rich genetic diversity for traits of economic value became subsequently a potential gene pool for directed improvement of the crop. In this endeavour, the first official rice variety GEB 24, a spontaneous mutant from popular landrace Konamani was released in 1921 from Coimbatore, which had served as the starting material for rice breeding in India and also at International Rice Research Institute (IRRI), apart from Latisail another Indian variety which has been used as a parent material of innumerable semi-dwarf high yielding *indica* varieties released in India and also globally through IRRI breeding lines.

Some of the popular pure line selections are MTU 1, MTU 15 and HR 19 of Andhra Pradesh; Chin 7 of West Bengal; Kolamba strains of Maharashtra; GEB 24, CO 25, ASD 1 of old Madras Province; T 141 and SR 26B of Odisha; Basmati 370 of Punjab and T 1366 of Uttar Pradesh. Breeding for adverse growing conditions had also led to the development of varieties *viz.*, ARC 353-648 of Assam; DWP-1311 of Andhra Pradesh; ADT 17, ADT 28, TNR 1, TNR 2 of Tamil Nadu, PTB 15 and PTB 16 of Kerala; FR 13A and FR 43B of Odisha; Hybrid 84 of West Bengal for deep water and flood prone areas; Kalarata 1-24 and Bhurarata 4-10 of Bombay Province; SR 26B of Odisha; Chin 13 and Chin 19 of West Bengal for saline conditions and AKP 1, AKP 3, BCP 2 and BCP 5 of Andhra Pradesh; ADT 22, ADT 24 of Tamil Nadu, PTB 18 of Kerala; BAM 15 of Odisha; N 22 of Uttar Pradesh; Chin 25 and Chin 27 of West Bengal for drought prone situations. Resistance breeding against biotic stresses led similarly to the development of CO 25, ADT 25 against blast and MTU 15, TKM 6, SLO 12 and CH 47 against stem borer (Siddiq and Viraktamath, 2000).

ADT 27, popularly called Kuruvai sirumani released in Tamil Nadu during 1964-65 was the only outcome of *indica japonica* hybridization in our country. The variety was derived from Norin 8/GEB 24 and matured in 105 days, which made double cropping possible in Cauvery Delta. This fertilizer responsive variety yielded on an average 4.8 t/ha.

Indonesia was the first country where the concept of multi location testing was executed to reduce the number of varieties under evaluation test. Its successful experience prompted breeders in many countries including India to follow and realize the need for improving the method of varietal testing. The use of statistical parameters *viz.,* variance and standard error helped to improve field testing technique, application of the principles of randomization and analysis of variance that decide on field plot design helped in the evaluation of entries more accurately.

Breeding of New Plant Types

On the strength of their long experience and experimental findings, breeders believed that genetic yield level of rice under irrigated ecologies could be raised by 20-25 per cent through enhancement of biomass per unit area without altering the current level of harvest *index* (Peng *et al.,* 2008) The strategy could be achieved by planting less profusely tillering but high panicle weight genotypes at high density by manipulation of crop geometry. The new plant type tailored in rice is of semi dwarf (~100 cm) stature with 7-8 productive and synchronized tillers, very strong culm with upright foliage, heavy panicles with more grain number (200-250 grains/panicle) and high test grain weight (~25g/1000 grains) (Plate 4.8). But the grain yield realized was low due to less biomass production and poor grain filling.

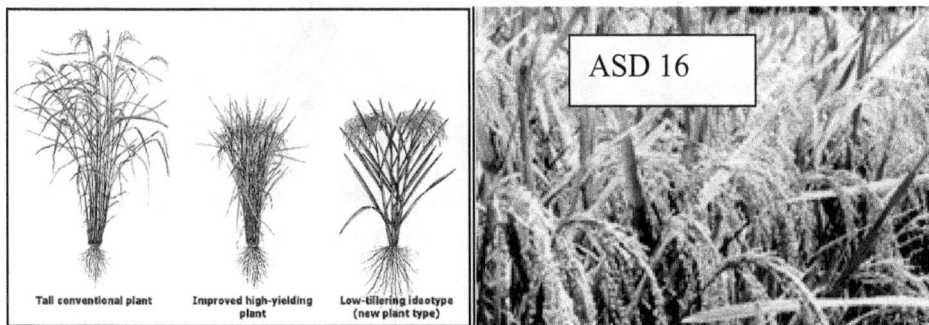

Tall conventional plant   Improved high-yielding plant   Low-tillering ideotype (new plant type)

ASD 16

Plate 4.8: New Plant Type.

Hence, they were not released for cultivation. In Tamil Nadu, a variety ASD 16 with new plant type characters *viz.*, erect, semidwarf, non lodging habit, compact panicles with around 250 – 300 grains of short bold nature with an average yield of 5.6 t/ha was released during 1986 from Ambasamudram, TNAU.

In 1995, breeding of second generation New Plant Types (NPT) lines began by crossing first generation tropical *japonica* NPT lines with elite *indica* parents to improve biomass production by increasing the tiller number and improving grain setting by reducing slightly the panicle size without change in panicle length and eliminating the compact arrangement of spikelets. This also helped to have pest and disease resistance and improved grain quality. Few second generation NPT lines registered significantly higher yield than *indica* check variety, IR 72. A second generation NPT line (IR 72967-12-2-3) produced 10.16 t/ha with significantly higher yield than the check variety PSBRc 52, mainly due to higher biomass and greater grain weight. Stimulated by IRRI's NPT program, China established nationwide mega project on "super rice" in 1996 with an objective that "super rice" should yield at least 10 per cent more than the widely grown popular varieties with acceptable grain quality. The "super rice" varieties can be developed by breeding inbred or hybrid varieties. A "super hybrid rice" breeding programme was started in 1998 by Prof. Yuan Longping to combine ideotype approach with the use of inter sub specific heterosis (Yuan Longping, 1997) (Plate 4.9). As a result, two super rice hybrids *viz.*, Xieyou 9308 and Liangyoupeijiu were developed. Maximum yield produced by Xieyou 9308 was 12.23 t/ha and it out yielded the best hybrid check (Xieyou 63) by 17.5 per cent. Another hybrid, Liangyoupeijiu yielded 12.11 t/ha, which was 28.6 per cent higher than that of hybrid check, Shanyou 63 in the experimental plots.

Plate 4.9: Super Hybrid Rice.

Exploitation of Hybrid Vigour

India started exploiting the technology since mid-90s taking advantage of the IRRI developed male sterile line IR58025A, well suited to typical tropical conditions like ours and of the experience of China in hybrid breeding/seed production (Jiming

Li and Yuan Longping, 2010). With about one ton yield advantage, as many as 90 hybrids have been released so far, for general cultivation. This much progress has been possible largely because of the active participation of the private sector which accounts for 70 per cent of the hybrid now planted commercially and 90 per cent of the hybrid seed produced and marketed. In the 18 years period since the introduction of rice, we could bring only two million hectares under hybrid rice. This remains in stark contrast to what had been the case in China, wherein less than the comparable period, over 13 million ha could be planted to hybrid rice. The slow pace of adoption is due to unsatisfactory and inconsistent yield advantage, lack of medium/medium late hybrids suited to long monsoon season in the traditional rice areas, susceptibility to major pests and less acceptable cooking quality. The popular hybrids include PA 6129, DRRH 2, Sahayadri 4 in the early duration group, Pusa RH 10, PA 6201, US 312, DRH 775, GK 503 in the mid early group and PHB 71, JKRH 401, HR I157, KRH 2, DRRH 3, CRHR 32, CORH 4 in the medium maturity group.

Breeding for Resistance to Biotic and Abiotic Stresses

Since the introduction of high yield technology which facilitated year round cropping, the spectrum of insect pests and diseases has widened from a few to many, causing serious crop losses year after year. Endlessly emerging virulent/viruliferous strains, especially in the absence of matching resistance gene(s) make the crop variety vulnerable to the pests. By stacking strain(s)-specific resistance genes, hitherto available against most of them by conventional backcross breeding or molecular marker assisted backcross breeding, desired level of broad spectrum resistance to the prevailing pests and their biotypes/pathotypes could be provided (Zhang, F and J. Xie. 2014). As many as 38 genes are known to provide resistance to BLB and its pathotypes, many are now linked to robust/gene markers. Similarly, many race specific resistance genes against blast disease, brown plant hoppers and gall midge have been linked to reliably usable markers. Using such trait linked markers, popular high yielding varieties/parental lines of ruling hybrids (Pusa Basmati 1, Samba Mahsuri and Pusa RH 10, IR 64 etc.) could be selectively introgressed with resistance genes $Xa31$, $xa13$ and $xa5$, $xa33$, $xa38$ against blight and *pitu* (t), $Pi38$, Pi^{kh}, $Pi48$, $Pi2$ against blast. Among the major insect pests, at least five resistance genes each against brown plant hopper (*Bph1*, *Bph2*, *Bph* 10 (t) and *Bph* 8) and gall midge (*Gm1*, *Gm2*, *Gm3*, *Gm4*, *Gm5*, *Gm7*) could be introgressed into some of the popular varieties by marker assisted breeding.

Among the abiotic stresses that severely depress yield, drought, submergence and salinity are most important stresses. Plant's ability to withstand such stresses is derived from cumulative network of physical and biochemical functions. Though sources of tolerance to all such stresses are available in the germplasm, progress to desired extent could not be achieved so far, since tolerance to abiotic stresses is genetically complex and polygenically controlled. As a result, over 55 per cent of rice area under rainfed uplands and lowlands could only be slowly improved for higher productivity.

Regarding drought, several marker linked tolerance QTLs (osmotic adjustment, cell membrane stability, relative water content, root length/mass etc as well as yield *per se*) have been mapped to various chromosomes (12, 8, 1 etc) and are being used in breeding for drought tolerance. The first aerobic rice variety Sahbhagi dhan (IR 74371-70-1-1-CRR-1), has been recently developed by using the yield specific QTL (DTY12.1) on chromosome 12 by IRRI. Similarly tolerance to *sub*1A (chr.9) accessed from the traditional flood resistant variety FR13A could be successfully used in developing varieties *viz.*, Swarna *sub*1A, Savithri *sub*1A, IR 64 *sub*1A, Samba mahsuri *sub*1A etc. Salt tolerance QTL (*saltol*) accessed from the salt tolerance Indian variety Pokkali is being introgressed into many varieties by marker assisted breeding.

Enrichment of Nutritional Quality

Rice is the staple food for more than half of the world's population and serves as a major source of dietary energy, protein, vitamins and essential fatty acids and micronutrients for the poor and vulnerable society in the world. It occupies the enviable prime place among the food crops cultivated around the world, yet it contains insufficient levels of the key micronutrients iron, zinc and vitamin A to meet the daily dietary requirements (Ravindra Babu, 2013). Even a small increase in its nutritive value could be highly beneficial for human health. Developing micronutrient dense rice, with higher amount of iron, zinc and vitamin A could have a beneficial effect on the health of low income people. Using plant breeding approach to address micronutrient malnutrition would provide a new tool to use in combating diet-related diseases. Identification of germplasm with high grain iron and zinc and understanding the genetic basis of their accumulation are under way. Several high yielding lines are being enriched with iron and zinc content. A line derived from Samba Mahsuri x Chittimuthyalu with short bold grains, semi-dwarf with high yield potential, medium duration, good quality characters with high iron and zinc in brown rice was identified. Transgenic efforts are positively underway to enhance the iron and zinc content in the endosperm of rice grains.

Future Demand Projections and Technological Opportunities

Of the two major cereals - wheat and rice, that contribute over 80 per cent to the country's food grain production, the latter's share is more than 43 per cent with a potential to raise further its contribution. Based on the consumption trend, it has been estimated that rice requirement would be 130 and 190 million tonnes respectively by 2025 and 2050, to sustain the current level of supply. Achievement of the projected demands in the face of many constraints would be a challenging task. In the era of genomics, solutions can be found to such constraints (Das, 2013). Among the technological opportunities to meet the medium term demand projection, sustenance and extensive adoption of hybrid technology, prospects of exploiting new plant type varieties now in the advanced stages of development and Green super rice could be of value. Also, the search for and use of still underexploited yield genes, tailoring new plant architecture, manipulation of biosynthetic pathway of starch and prospects of improving carbon assimilation to cope with changing climate are contemplated as means to achieve long term production demands.

References

Chang T.T. 1976. The origin, evolution, cultivation, dissemination and diversification of Asian and African rices. *Euphytica* 25: 431-441.

Das, S.R. 2012. Rice in Odisha, IRRI Technical Bulletin No. 16 Los Banos (Philippines): International Rice Research Institute. 31 p.

Das, S.R. 2013. Reflections on > 40 years of rice breeding for eastern India. *SABRAO Journal of Breeding and Genetics*. 45 (1) 21-26.

Harlan, J.R and J.M.J. Dewet, 1971. Towrds a rational classification of cultivated plants. *Taxon*, 20: 509-517

Hung, Y. 2001. Rice ideotype breeding of Guangding Academy of Agricultural Sciences in retrospect. *Guanf donf Agric. Sci*. 3, 2-6.

Jiming Li and Yuan Longping. 2010. Hybrid Rice: Genetics, Breeding, and Seed Production. *Plant Breeding Reviews*, 17: 15–158,

Molina, J., Sikora, M., Garud, N., Flowers, J.M., Rubinstein, S., Reynolds, A., Huang, P., Jackson, S., Schaal, B.A., Bustamante,C.D., Boyko, A.R., Purugganan,M.D., 2011. Molecular evidence for a single evolutionary origin of domesticated rice. In: Proceedings of the National Academy of Sciences of the United States of America, www.pnas.org/cgi/doi/10.1073/pnas.1104686108

Peng. S, G. S. Khush, P. Virk, Q. Tang and Zou. Y. 2008. Progress in ideotype breeding to increase rice yield potential. *Field Crops Research*. 108(1): 32–38.

Ravindra Babu. V. 2013. Importance and advantages of rice biofortification with iron and zinc. *Journal of SAT Agricultural Research*. pp11

Saiato H, Y.Okumato, T.Teranishi, Y.Qungbo, T.Nabazaki and T.Tanisaka. 2007. Heading time genes responsible for regional adoptability of Tongiltype short cusumed rice cultivars developed in Korea. *Breed Sci*. 57: 135-143.

Siddiq, E.A and B.C. Viraktamath. 2000. Rice. In Breeding Field crops. Edited by V.L. Chopra, Oxford and IBH, New Delhi, pp.1-85

Tateoka, 1962. Taxonomic studies of *Oryza* L. *O. latifolia* complex Botanical Magazine 75, 418-427.

Yan, W., Q. Jin, F. Yu, K. Tomita and H. Hisamitsu. 2001. Comparison of nutrient quality between new and stored grains in japonica rice. International Rice Research Notes,26(2): 26

Yuan Longping. 1997. Hybrid rice breeding for super high yield in IRRI discussion paper series No.31 Editory by G.L. Denning and T.W. Mew. Proceedings of the China-IRRI dialogue held in Beijing, China: 7-8 November, 1997. pp.10

Zhang, F and J. Xie. 2014. Genes and QTLs Resistant to Biotic and Abiotic Stresses from Wild Rice and Their Applications in Cultivar Improvements in Rice-Germplasm, Genetics and improvement. pp.59-78

Chapter 5

Optimized Land Preparation for Rice Cultivation

N. Asoka Raja

Department of Agronomy, Tamil Nadu Agricultural University
Coimbatore – 641 003, Tamil Nadu
e-mail: agronomy@tnau.ac.in

Rice is one of the three most important food crops in the world and it is the staple food for over 2.7 billion people. Rice cultivation is the principal activity and source of income for millions of households around the globe. Several countries of Asia and Africa are highly dependent on rice as a source of foreign exchange earnings and government revenue. World top ten rice producing countries are China, India, Indonesia, Bangladesh, Vietnam, Thailand, Myanmar, Japan, Philippines and Brazil.

In India, area under rice is 44.6 m ha with total output of 80 million tonnes (paddy) with an average productivity of 1855 kg/ha. It is grown in almost all the states. West Bengal, Assam, Bihar, Uttar Pradesh, Punjab, Madhya Pradesh, Maharashtra, Odisha, Andhra Pradesh, Karnataka, Tamil Nadu and Kerala are major rice growing states and contribute to total 92 per cent of area and production.

Rice is grown in diverse ecosystem and being a semi aquatic plant, transplanting method of establishment is more favourable and in general assumed to give more staple yield. Transplanting is most common in low land (wet preparation of fields) where there is assured irrigation water for the crop growing period and is usually preceded by puddling.

Rice is grown in various system of cultivation throughout the world based on geography, land and water resources.

1. Transplanted puddled lowland rice.
2. Wet seeded puddled lowland rice.
3. Dry seeded rainfed unpuddled lowland rice.

4. Dry seeded rainfed unpuddled lowland rice (supplement irrigation).

5. Dry seeded irrigated unpuddled lowland rice.

I. Nursery Management

1. Wet Nursery

Nursery Area

☆ Select 20 cents (800 m²) of land area near to water source for transplanting one hectare in conventional method.

Nursery Bed Preparation

The area should have an assured water supply and an efficient drainage system. It should be dry ploughed twice with cultivator and apply 1 tonnes of FYM or compost to 20 cents nursery. Later, it should be irrigated and allowed to be wet for another two days. Afterwards it should be puddled twice and the puddling may be repeated after a gap of one week. Basal application of DAP 40 kg is recommended when the seedlings are to be pulled out in 20-25 days after sowing in less fertile nursery soils, and if not readily available, apply straight fertilizers 16 kg of urea and 120 kg of super phosphate. After levelling and final puddling, beds of convenient length (8-10 m) with width of 2.5 m are to be made, leaving 30 - 50 cm channels in between two beds. Sow the sprouted seeds uniformly on the seed bed.

2. System of Rice Intensification (SRI)

Mat nursery seedlings are established in a layer of soil mix, arranged on a firm surface (Concrete floor/polythene sheet/seedling trays). Seedlings are ready for planting within 14-20 days after seeding (DAS).

Nursery Area

Required nursery area is 100 m²/ha (or) 2.5 cent/ha - 1cent/acre upland.

Nursery Bed Preparation

Select a level area near the water source with efficient drainage system. The surface should be covered with banana leaves with the mid-rib removed or polyethylene sheets or any flexible material or cemented floors to prevent seedling roots from penetrating to the bottom soil layer.

Preparation of Soil Mixture

Four (4) m³ of soil mix is needed for each 100 m² of nursery. Mix 70 per cent soil + 20 per cent well-decomposed pressmud/compost/FYM + 10 per cent rice hull. Incorporate 1.5 kg of powdered DAP with the soil mixture.

Filling the Soil Mixture

Place a wooden frame of 0.5 m long, 1 m wide and 4 cm deep divided into 4 equal segments on the plastic sheet or banana leaves, fill the frame almost to the top with the soil mixture.

Main Field Preparation

☆ Puddled lowland prepared as described in transplanted rice.

☆ Perfect leveling is a pre - requisite for efficient water management.

Pre-germinating the Seeds

Soak the seeds for 24 hrs, drain and incubate the soaked seeds for 24 hrs, Sow when the seeds sprout and radicle (seed root) grows to 2-3 mm long and cover them with dry soil to a thickness of 5 mm.

3. Dry Nursery

Nursery Area

Area : 20 cents. This type of nursery is handy in times of delayed receipt of canal water or areas where sufficient water is not available. The field is dry ploughed 5-6 times to acquire the fine tilth. Nursery area with sand and loamy soil status is more suitable for this type of nursery. Plots of 1 to 1.5 m bed width and channels may be formed. Length may be according to the slope and soil. After preparing the beds they are to be wetted and 1 tonne of FYM, 2.5 kg of P_2O_5 and 2 kg of potash need to be applied. Sowing may be dry seeding. Seeds may be covered with sand and finely powdered farm yard manure.

Land Preparation–Transplanted Puddled Lowland Rice

Plough the land during summer to economize the water requirement for initial preparation of land. Flood the field 1 or 2 days before ploughing and allow water to soak in. Keep the surface of the field covered with water. Keep water to a depth of 2.5 cm at the time of puddling.

II. Main Field Preparation

1. Puddling

Rice growth and yield are higher when grown under submerged conditions. Maintaining standing water throughout the crop period is not possible without puddling. Puddling is ploughing the land with standing water so as to create an impervious layer below the surface to reduce deep percolation losses of water and to provide soft seedbed for planting rice.

Puddling operation consists of ploughing repeatedly in standing water until the soil becomes soft and muddy. Initially, 5-10 cm of water is applied depending on the water status of the soil to bring it to saturation and above and the first ploughing is carried out. After three to four days, another 5 cm of water is applied and after 2-3 days second ploughing is carried out. By this operation, most of the clods are crushed and majority of the weeds are incorporated. Within 3-4 days, another 5 cm of water is given and third ploughing is done in both the directions. The third ploughing can be done either with a wetland plough or with a wetland puddler (Tractor drawn cage wheel).

Planking or leveling board is run to level the field. To know whether puddling is thorough or not, a handful of mud is taken into the hand and pressed. If it flows freely through fingers and if there are no hard lumps, puddling is considered to be thorough. Unlike in other tillage operations, puddling aims at destroying soil structure. The individual soil particles *viz.*, sand, silt and clay are separated during puddling operation. The soil layer with high moisture below the plough sole is compacted due to the weight of the plough. The soil particles separated during puddling will settle later. The sand particles reach the bottom, over which silt particle settle and finally clay particles fill the pores thus making impervious layer over the compacted soil.

Puddling is done with several implements depending on the availability of equipments and nature of the land. Soils with bulk density of less than 1.0 are considered as problem soils as puddling with animal drawn implements is difficult. The feet of the animal sink very deep during puddling. Under such situation, puddling is done with spades by manual labour. Most of the farmers use wetland plough or mould board plough.

Wetland puddler consists of a series of blades attached to a beam at an angle. When it is worked, the soil is turned and puddling operation is completed quickly compared to the country plough. Generally, green manure is applied to rice field which is incorporated by green manure trampler. Tractor drawn implements can be used for puddling by attaching cage wheel to prevent sinking.

2. Wet Seeded Puddled Lowland Rice

On receipt of showers during the months of May - July repeated ploughing should be carried out so as to conserve the moisture, destroy the weeds and break the clods. After inundation puddling is to be done as described earlier. More care should be taken to level the field to zero level. Stagnation of water in patches during

germination and early establishment of the crop leads to uneven crop stand. Land leveling contributes for more efficient weed and water management in mainfield. Provision of shallow trenches (15 cm width) at an interval of 3 m all along the field will facilitate the draining of excess water at the early growth stage.

3. Dry Seeded Rainfed Un-puddled Lowland Rice

Dry plough to get fine tilth taking advantage of rains and soil moisture availability. Apply gypsum at 1 t/ha basally wherever soil crusting and soil hardening problem exist. Perfect land leveling is to be done for efficient weed and water management. Provide shallow trenches (15 cm width) at an interval of 3m all along the field to facilitate draining excess water at the early growth stage.

4. Dry Seeded Rainfed Un-puddled Lowlad Rice (With Supplemental Irrigation)

Dry plough to get fine tilth taking advantage of rains and soil moisture availability. Apply gypsum at 1 t/ha basally wherever soil crusting and soil hardening problem exist. Perfect land leveling is to be done for efficient weed and water management. Provide shallow trenches (15 cm width) at an interval of 3m all along the field to facilitate draining excess water at the early growth stage.

5. Dry Seeded Irrigated Un-puddled Lowland Rice and Dry Seeded Upland Rice

Dry plough to get fine tilth taking advantage of rains and soil moisture availability. Apply gypsum at 1 t/ha basally wherever soil crusting and soil hardening problem exist. Perfect land leveling is to be done for efficient weed and water management. Provide shallow trenches (15 cm width) at an interval of 3m all along the field to facilitate draining excess water at the early growth stage.

Chapter 6

Rice Nursery: Types, Practices and Issues

R. Rajendran, E. Subramanian and K. Subrahmaniyan

Tamil Nadu Rice Research Institute,
Aduthurai – 612 101, Tamil Nadu
e-mail: rajendrankmu@yahoo.co.in; dirtrri@tnau.ac.in

Rice is cultivated under various eco systems, broadly classified based on source of water used as rainfed or irrigated. Based on water and land use, rice farming is further classified as lowland (bunded with stagnation of water) and upland (unbunded, no standing water; instead, freely allowed to move down according to the topography). About 50 per cent of the rice area in Asia is under low land transplanted culture. Under this system, land is either prepared wet or dry but water is stagnated in the field by bunds. About 30 per cent of the world's rice area is under rainfed lowland condition where rain water is stagnated by bunding. For both the bunded rice cultivation methods (irrigated/rainfed lowland), seedings are transplanted by raising in the nursery. Rice, which is to be transplanted into lowland puddled soil must first be nursed on seedbeds. The main reason for nursing rice is simple: to give the seedlings a substantial head start on weeds.

The nursery management, handling of seedlings and plant density are more or less similar for both these type of rice culture. In case of rainfed lowland rice culture in the tropics, seedlings are grown in a season during which adequate water supply/moisture is assured, mainly coincide with seasonal rainfall periods. In the case of irrigated lowland rice, seedlings are mostly transplanted throughout the year as and when needed, because of assured water supply through canal/tanks/ground water.

Types of Rice Nursery

Rice nursery raising is traditionally followed by farmers in various ways. Nursery is raised where there is assured and adequate supply of water available.

Though various types of rice nurseries are available in different states/countries, the following three major nursery management methods are predominantly practiced by farmers rice growing countries of Asia including Tamil Nadu in India depending upon the water and labour availability and method of rice transplanting (manual/machine).

1. Wet-bed nursery – traditional manual transplanting
2. Dry-bed nursery – manual transplanting
3. Modified mat nursery - for SRI, machine transplanting and seedling throwing

1. Wet-bed Nursery

Traditional wet-bed nurseries are preferred under irrigated systems. In the wet-bed nursery, sprouted seeds are sown on puddled soil after leveling the nursery The rice seedlings are transplanted in the main field manually. Rice seeds are soaked in water for 24 hours and then incubated for 24-36 hours to get pre – germinated (sprouted) seeds. The duration of incubation varies depending upon the stage of sprouted seed required. Farmers use "Irandam Kombu" or "Munam Kombu" seeds which indicates the state of sprouted seed with varying plumule length. Farmers mostly use the seed at "Irandam Kombu" stage to avoid damage to sprouted seeds while handling/sowing. This pre-germination process, assures a quick and uniform start of the seedbed in the nursery. The success of the transplanted rice and bumper harvest depends on the quality of seedlings raised in the nursery. The seedlings should be robust and healthy for better establishment, early vigor, growth to attain the targeted grain yield.

In the wet-bed method, pre-germinated seeds are broadcast uniformly on a raised bed and the seedlings are ready for transplanting at 15 - 35 days after sowing. The age of seedlings for transplanting varies with duration of the variety chosen for cultivation. A general thumb rule for seedling age is, one week nursery duration for one month duration of a cultivar. Scientifically, seedlings at 3rd - 4th leaf stage before tillering are considered as ideal seedling for transplanting for getting better growth and yield. Rate of seeding varies with grain size (14 - 24 g test weight) and duration of the variety grown (100 - 160 days). Rate of seeding is about 37.5 - 75 g per m^2 of seedbed, which amounts to a rate of about 30 - 60 kg/ha. In Tamil Nadu, 800 m^2 nursery is normally recommended, but practically reduced nursery area is often used by the farmers with congested and thick seed rate. The seedlings do not attain the expected growth in many occasions for which higher seed rate is the main reason among other factors. Apart from the poor nursery fertility and fertilization, thick sowing of seed results in puny and unhealthy seedlings which led to transplanting of more number of seedlings per hill and poor establishment. In situ green manure incorporation, proper land preparation with good leveling, optimal seed use, seed treatment with bio-fertilizer/pesticides, efficient water management are the important agronomic practices to be followed for getting robust and healthy seedlings from traditional wet-bed nursery. Seed treatment with bio - control agents (_Pseudomonas fluorescens_ 10 g/kg of seed), bio - fertilizer (_Azospirillum_), fungicides

Wet-bed Nursery–Seeds Sowing

Wet-bed Nursery–Pulling Out the Seedlings

(carbendazim/tricyclozole), nutrient solutions (1 per cent KCl) are recommended depending upon the need. Tamil Nadu Agricultural University revised the DAP application at 10-15 days prior to pulling out of seedlings instead of basal application at the time of sowing. Basal application of DAP encourages deeper root system which results in cumbersome pulling out and labour drudgery besides snapping of seedlings especially when pulling out of seedlings is delayed for want of sufficient water or labour. This type of wet-bed nursery is a labour intensive method makes the puuling out and transport operations more cumbersome and costly besides consuming more water and seed rate.

In recent years poor attention is paid on nursery management resulting in poor use efficiency of precious agricultural inputs *viz.*, seed, water and fertilizers. Labour scarcity and high wage rates make the farmers to lose interest in rice farming. Hence, there is an urgent need to increase the awareness of farmers on innovative and improved nursery management technologies in order to conserve the natural resources of seed, water and labour.

2. Dry-bed Nursery

The dry-bed method is not extensively practiced in the Asian tropics, although it is popular in the Philippines on clayey soils in the rainfed lowland areas where there is insufficient water to irrigate seedbeds. Dry-bed seedlings are grown in a manner similar to that of wet-bed seedlings, except that the soil is not puddled and drainage is provided to keep the soil moist but not inundated. Land is prepared dry by breaking clods using hand or stick and levelled. In this method, soaked seeds are spread over seedbed and then irrigated. In some areas of Punjab dry seeds are sown under dry condition in shallow depth, covered with thin layer of sand and finely powdered FYM. The seeding rate and other agronomic practices are same as followed for wet bed nursery. The seed beds need to be covered with coconut fronds or polythene sheet to avoid the damage of emerging seedlings by heavy rains and birds. The beds should be watered thoroughly, immediately after planting and twice a day thereafter. If irrigation water is available, water can be taken along the channels and splashed onto the beds. If the seed-beds dry out for even one day, the growth of the seedling will be seriously impaired. In the Philippines, farmers construct a small pond near a source of water and water the seedlings two or three times daily. In Cauvery delta region and few parts of Tamil Nadu also, dry bed nursery is practiced still, mainly because of water scarcity. Light soil with sandy status is more suitable for this type of nursery. Farmers could opt for this type of dry-bed nursery, provided if they have assured irrigation source to water regularly.

Dry-bed Nursery–Layout   **Dry-bed Nursery**

Dry-bed seedlings will not grow as fast as wet-bed seedlings. Seedlings on the dry bed should be ready for transplanting from 25 days onwards.

3. Mat Nursery

a. The *Dapog* Nursery

The *dapog* method of raising seedlings originated in the Philippines and is now, fairly common in South and Southeast Asia. The *dapog* nursery is constructed to raise seedlings without any soil whatsoever. Rice seeds contain sufficient food in the endosperm to permit the young seedlings to grow up to 14 days without receiving any external nutrients except air, water, and sunlight (Patel *et al.*, 2011). Consequently, it is possible to nurse seedlings without actually sowing them in soil. Seed-beds of approximately 1 m wide and 10-20 m long are laid out with a layer of plastic sheets or banana leaves. The walls of the bed are raised with bamboo split

Dapog Nursery   **Dapog Nursery–Seedling Mat**

stakes. In this method, very high seed rate is used which germinates and form a mat like layer of adventitious roots, called as *suruttuppai/pai/dapog* nursery. Pre - germinated seeds @ 1 kg/m^{2}are sown directly on the plastic sheets or banana leaves and about 60 kg of seeds will be required to plant one hectare of field. Seed beds are covered with dry rice straw to protect the seedlings from birds and the straw can be removed on sixth day when the seedlings are large enough that birds will no longer eat them.

The *dapog* nursery can be located anywhere convenient, as long as it is near a reliable water supply and where it can be watched at all times, since they require constant watering and protection from birds damage. The seedlings grown in this type of nursery are very short, delicate and thin which led to planting more number of seedlings (8-12) per hill by machine or labour. Because of the difficulties faced in handling of rice seedlings and poor crop establishment, *dapog* nursery did not pick up in farmers' field.

b. Modified Rice Mat Nursery

Rice crop establishment is changing its dimension in recent years due to reasons of water and labour scarcity. Acute labour shortage and drudgery faced by human labour while transplanting force the farmers to go in for mechanized transplanting. In the Cauvery delta region bwhich is known as *rice bowl* of Tamil Nadu, mechanized transplantation is fast picking up now. Introduction innovative crop establishment methods in the name of system of rice intensification (SRI) has created more awareness on adoption of mat nursery system in Andhra Pradesh, Tamil Nadu and Karnataka. Among the agronomic practices recommended for SRI method of planting, transplanting single and young seedling at one per hill warranted the production of robust and healthy seedlings in about 15 days. Under the SRI method, 8 – 15 days old seedlings are manually planted in the main field. A modified rice mat nursery was developed and released by Tamil Nadu Agricultural University in 2006 which could give 18-20 cm tall robust seedlings in 15 days time and found suitable for SRI method of planting (Rajendran *et al.*, 1995). Labour scarcity force the farmers to go in for rice crop establishment my machines. Machine transplanting also requires rice mat nursery, but slightly different from the mat nursery required for SRI method of rice planting. A mat nursery for SRI need 4 cm high seed bed with reduced seed rate of 5-7 kg ha^{-1}, where as the mat nursery indented for machine planting require a thin seed bed in readymade trays and a high seed rate of 25-40 kg ha^{-1}.

i. Modified Mat Nursery for SRI

The modified rice mat nursery is nothing but an improvement over the age old *dapog* nursery. In the modified mat nursery, a thin layer of seed bed with rooting medium is provided to get robust and healthy seedlings fit for transplanting at young age (14 – 18 days). Following are the management practices, identified as ideal for production of robust and healthy seedlings suitable for SRI method where seedlings are manually transplanted.

☆ Shallow raised bed (5-10 cm height) with one meter width and convenient length; 100 m² nursery bed is required for planting one hectare. Raised bed is surrounded by narrow channels to retain water (after 5th day) and bunds all around leaving 30 cm walking space between 2 beds.

☆ Seed beds are covered by perforated 300 gauge polythene sheet to prevent roots growing deep but facilitating drainage of water underneath.

☆ Seed bed medium prepared with native soil and well decomposed FYM (1: 10 ratio) or mixed with vermin compost/well cured pressmud. Native soil itself is enough if found highly fertile; medium is poured into the seed frame having a height of 4 cm, compartmentalized in to 4 boxes each having 0.125 m² area so that one m² of seed frame accommodates 8 seed boxes. Seed bed medium is mixed with powdered DAP @ 1 kg/ha of nursery bed, approximately 25 g/m² seed bed.

☆ Seed rate for all varieties ranges from 5 – 7 kg/ha which works out to 50 g/m² seed bed. Adoption of thin seed rate is essential to get robust seedlings.

☆ Seed beds are covered with well thrashed paddy straw to avoid admixture of seed or dried coconut fronds with little aeration up to 5 days. This will promote germination due to induced temperature.

☆ Sprinkling water using rose can up to five days morning and evening, first watering should be a soaking irrigation to facilitate easy and quick germination. Water should be let in water all around the beds after 5 days. Seed beds should not be inundated/flooded with standing water as it will affect and stunt the seedling growth.

☆ Mixing *Pseudomonas fluorescence and drenching the same by* mixing with water @ 25 g/m² seed bed on 6th day gives robust growth of seedlings.

☆ If necessary drenching seed *beds with* 0.5 per cent urea solution on 9th day after sowing.

☆ Seedlings can be lifted as cakes and easily transported to main field for planting at 14 – 15 days or when seedlings attain 4 leaf stage.

Modified Mat Nursery   **Modified Mat Nursery–Seedling Mat**

Modified rice mat nursery saves precious farm resources and inputs *viz.*, reduce nursery area (90 per cent), seed (80 per cent), (water 50 per cent), fertilizer (90 per cent) and labour (35 per cent). In this method robust rice seedlings of 18-20 cm tall with 4-5 leaves are obtained on 15th day. The best age for transplanting seedlings is 15 days but can be kept in mat nursery up to a maximum of 20 days beyond which seedling would turn yellow and dry up as seedlings are grown on thin layer of seed bed medium only, unlike the traditional wet-bed nursery method.

ii. Tray Nursery for Machine Transplanting

Unlike the modified rice mat nursery of SRI method, the tray nursery meant for machine transplanting require thin seed bed medium (1-2 cm), higher seed rate (25 - 40 kg/ha) preferably on HDPE or LDPE plastic trays having desired size of individual compartments or cakes suiting to the type of transplanter used for planting. This is the essential pre – requisite for machine planting as the seedling cakes should fit in to the transplanter chambers tightly to facilitate the planting fingers for picking up rice seedlings rightly.

Tray Nursery   **Machine Transplanting**

On an average, 36 - 40 m² nursery bed is required to transplant one hectare by machine ie about 200 - 220 tray seedling are required to plant one hectare. Other management practices are same as that of modified rice mat nursery of SRI method of planting. For machine transplanting, 15 - 18 days old seedlings are used and can be kept up to 20 days beyond which the seedlings started yellowing and drying due to thin seed bed trays and high seed rate used. Planting of too many seedlings by the machine (5-6) per hill is the serious field problem faced by the farmers and private entrepreneurs. The transplanting machine itself has minor adjustments to pick up fewer seedlings, however still the issue of planting more number of seedlings per hill still exists at field level. This is mainly due to use of higher seed rate leading to thin and delicate seedlings produced in the tray nursery. Of course, based on the lessons learnt and farmers' feedback, private nursery entrepreneurs reduced the seed rate from 50 kg/ha to 25 kg/ha. Now a day, seeding machines are available for placing seeds sparsely to facilitate the transplanting machine to pick up 2-3 seedlings per hill. Construction of precisely sown tray nursery and transplanting single seedling per hill would be a significant achievement as it could pave way to manage the labour scarcity and get bumper rice yields as in the case of SRI in future.

iii. Bubble Tray Nursery

Seedlings are raised on plastic trays of convenient size embedded with miniature cups looking like bubbles from bottom side and each bubble cup can

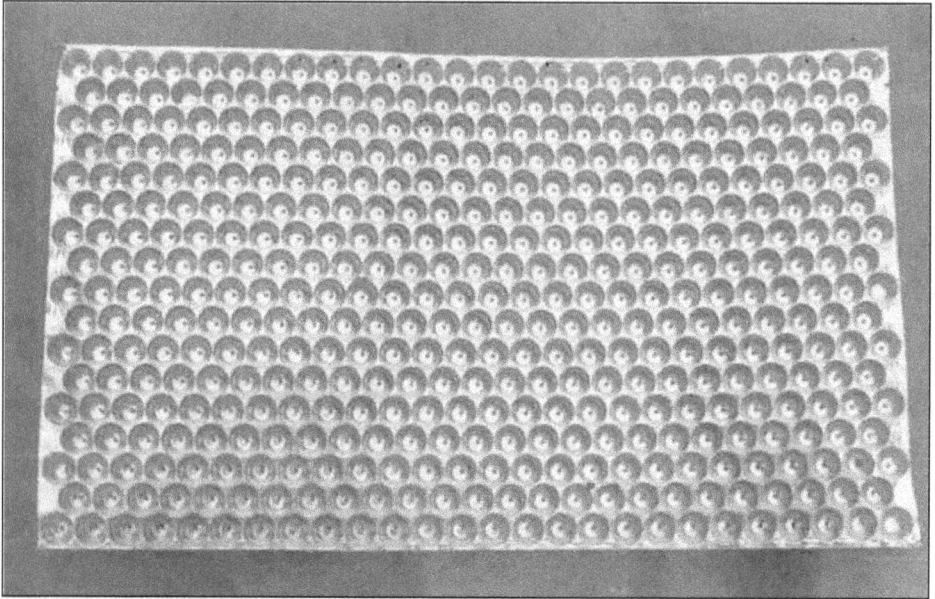

Bubble Tray

Bubble Tray Nursery–Seedlings

hold 2-3 seeds. About 250 m² seed bed area holding approximately 750 trays will be required to plant one hectare. Nursery trays can be kept nearer to water source either in upland/lowland or near the house. Seedlings with a root ball of earth will be ready in 12 -15-days time suitable for throwing method of rice crop establishment. The ball of earth adhering to the seedlings is making it easy for thowing the seedlings at targeted location and achieving a good contact with field soil. Tamil Nadu Agricultural University studied the seedling throwing method using the seedlings raised from the traditional nursery itself and came out with positive results in Cauvery delta research centres at Aduthurai and Thanjavur. About two third of labourers could be saved when compared to the traditional transplanting method. However, seedling throwing method is suitable only for dry seasons and fields having adequate drainage facility (Sanbagavalli and Kandasamy, 2000).

Table 6.1: Rice Nurseries–Summary

Nursery Type	To Plant One ha of Main Field		Optimum Seedling Age (Days)	Crop Establishment Method
	Nursery Area (m²)	Seed Rate (kg/ha)		
Wet-bed	800	30–60	20–35	Manual
Dry-bed	500	50	25–30	Manual
Dapog	60–75	40–50	9–14	Machine/manual
Modified mat	100	5–7	15	SRI
Tray	36–40	25–40	18–20	Machine
Bubble tray	250	15–20	12–15	Seedling throwing

In all the above innovative methods of rice nursery, owing to reduced nursery area, the cost of plant protection is not prohibitive. The choice of seedbed method depends on the availability of water. Studies conducted in India showed similar growth characteristics and grain yield with seedlings raised by above different mat nursery methods. Taller seedlings are desirable in the wet season (rainy) because of sudden floods, which many cause damage to the transplanted rice crop during early stages.

Community Rice Mat Nursery

Rice farmers lose interest to continue rice farming due to complex production situation prevailing. In addition to water scarcity and labour cost, escalation of other input costs and stagnating rice procurement prices reduce the profit of rice farming. Technologies have to developed and promoted to reduce the cost of cultivation in order to make the rice farming more profitable in line with other commercial crops grown. Rice nursery management is one area not yet fully explored which could bring down the rice production cost especially when farmers take up modified rice mat nursery in a community approach. Under this community system, rice mat nursery saves precious water apart from other benefits of sharing labour and reducing seed and fertilizer rates. On an average, modified rice mat nursery saves about 50 per cent of cost as compared to traditional wet nursery. This type of community nursery can very well be followed for conventional nursery method also. But the benefits will be manifold when farmers resort to community mat nursery approach. The private companies charge heavily for rice mat nursery for machine transplanting, instead if the farmers are trained on community system of mat nursery, considerable amount of money could be saved by the farmers. Transplanting alone can be done by hiring of machineries which also could be done by farmer groups collectively for which the government is now supplying the transplanters to farmer groups to promote custom hiring practice. Field level extension functionaries and contact farmers need to be trained on how to raise the modified rice mat nursery as this type of nursery could save considerable amount of rice production cost apart from judicious use of precious farm inputs *viz.,* seed and water.

References

Patel, D.P., A. Das, G.C. Munda, P.K.Ghosh, S.V. Ngchan, R.Kumar and R.Saha. 2011. Preparation of Modified Mat Nurseries (MMN) for Improved Rice Seedling Production. A Regional Supplement to ECHO Development Notes, ECHO Asia Notes, No. 9. ICAR Research Complex for NEH Region, Umiam, Meghalaya, India.

Rajendran, R., Ravi, V., Valliappan, K., Nadanasabapathy, T., Jayaraj, T., Ramanathan, S. and V. Balasubramanian, 2005. Early production of robust seedlings through modified mat nursery for enhancing rice (*Oryza sativa*) productivity and profit. *Indian J. Agron.* 50(2): 132-136.

Sanbagavalli, S. and Kandasamy, O.S. 2000. Nitrogen management and economic returns of seedling throwing method of rice planting in dry season. *Agricultural Science Digest*, 20(1): 42-45.

Chapter 7

Nutrient Management in Rice

P. Stalin and R. Jayakumar

Department of Soil Science and Agricultural Chemistry,
TNAU, Coimbatore, Tamil Nadu
e-mail: stalinpreemon@yahoo.com

Present Scenario

Rice is the 'Global Grain' and a staple food for about 50 per cent of the world's population that resides in Asia, where 90 per cent of the world's rice is grown and consumed. Within Southeast Asia, rice provides about 60 per cent of the human food consumption. About 55 per cent of the Asian rice is produced in irrigated areas, which accounts for about 75 per cent of Asia's total rice production with an estimated 2.2 billion Asian rice farmers and consumers depending upon the sustainable productivity of the irrigated lowland rice ecosystem for their food supply (Buresh *et al.*, 2005). The yield levels in India are low at 2.04 tonnes per ha compared to other major rice producing countries such as Japan (6.52 t/ha), China (6.24 t/ha) and Indonesia (4.25 t/ha). India would need to produce 143 million tonnes of rice to meet the growing population by 2030 (Subbaiah *et al.*, 2001). This increase in production could be achieved by intensification of rice cultivation rather than increasing the area. With the intensification of paddy cultivation, the soils have been depleted of several plant utrients including micronutrients. In Tamil Nadu rice is cultivated in 1.904 M ha with the production of 7.46 million tones and productivity of 3.92 t ha^{-1} (Agriculture Statistics, 2011-12).

Rice Based Cropping System

Different cropping systems are being followed in the different zones of Tamil Nadu based on the soil type, rainfall pattern, availability of irrigation water etc. The rice based cropping systems and rice involving crop rotations followed in Tamil Nadu are furnished below:

Rice-Based Cropping System

☆ Rice – Rice - Fallow

☆ Rice - Rice - Rice fallow pulses

☆ Rice – Rice - Cotton

☆ Rice - Rice - Gingelly

☆ Rice - Groundnut

Other Cropping System

☆ Sugarcane - Rice – Banana

☆ Turmeric - Rice

Systems of Rice Cultivation in Tamil Nadu

The various systems of rice cultivation in Tamil Nadu include

a. *Transplanted*: In all the Districts of TN

b. *Direct wet seeded*: In all the Districts of TN

c. *Direct dry seeded*: Kanchipuram and Tiruvallur (sl soils), Pudukottai, Nagapattinam and Thiruvarur(clay)

d. *Semi dry*: Kanchipuram and Tiruvallur, Ramnad, Tirunelveli, Tuticorin, Sivagangai,Kanyakumari Pudukottai, Nagai, Madurai, Dindigul, Theni and Thiruvarur

e. *Rainfed*: Kanchipuram, Tiruvallur, Pudukottai, Ramnad, Virudhunagar, Sivagangai and Kanyakumari

Rice Growing Seasons

Based on the cropping systems practiced in the different zones, rice is being grown in the following crop seasons in Tamil Nadu:

Rice Seasons of Tamil Nadu

☆ Sornavari	Apr - May
☆ Early Kar	Apr - May
☆ Kar	May – June
☆ Kuruvai	June - July
☆ Early Samba	July - Aug
☆ Samba	August
☆ Late Samba Thaladi/Pishanam	Sep – Oct
☆ Late Pishanam	Sep -Oct
☆ Late Thaladi	Oct - Nov
☆ Sornavari	Apr - May
☆ Early Kar	Apr - May

Duration Groups of Rice

Rice genotypes with varied duration to suit the different seasons and situations are being grown in Tamil Nadu. The most popular genotypes and their duration are furnished below:

Duration	Days	Varieties	Hybrids
Very early	< 100	CR 666, ADT 48	
Short	100 – 120	ADT 36, 37,42, 43, 45, and 47 CO 47, ASD 16, 18 and 20, TKM 9, IR 36 and 50, MDU 4 and 5. TKM 12, TRY 2, PMK 3, Anna 4	CORH1, CORH 3 ADTRH 1
Medium	121 – 140	ADT 38, 39, 46 and 49, IR 20, CO 43 and 46, ASD 18,19, Bhavani, I. White Ponni, TPS 2, TKM 10, TRY 1, CO 48, CO 49, CO 50	CORH2, CORH 4
Long	141 – 160	CO 40, ADT 44, ADT 50	
Very long	> 160	CR 1009	

Rice Varieties in for different Rice Cultivation Systems

a. Transplanted Rice and Wet Seeded Rice

Sornavari (April-Mar)

ADT 36, ADT 37, ASD 16, ASD 18, MDU 5, IR 50,ADT 43, CO 47, ADT (R) 45, ADT (R) 47, ADT (R) 48,CORH 3

Kar (May - Jun)

ADT 36, ASD 16, ASD 18, MDU 5, IR 50, ADT 43,CO 47, ADT (R) 45, ADT (R) 47, CORH 3

Kuruvai (Jun-Jul)

ADT 36, ADT 37, ASD 16, ASD 18, MDU 5, IR 50,ADT 43, ADT (R) 45, ADT (R) 47, ADT (R) 48, CO 47,CORH 3

Samba (Aug)

I.White Ponni, CO 43, CR 1009, ADT 38, ADT39, TRY 1*,TRY 3*, ADT (R) 44, ADT (R) 46, CORH 4, ASD 19,CO (R) 48, CO (R) 49,CO (R) 50, ADT (R) 49,TNAU Rice ADT 50, Swarna sub1, CR 1009, Bhavani, Paiyur 1

Late Samba

I.White Ponni, ADT 38, ADT 39, TRY 1*, TRY 3*, ADT 44, ADT (R) 46,ADT (R) 49, CO (R) 48, CO (R) 49, CO (R) 50, CORH 4, Bhavani

Navarai (Dec- Jan)

ADT 36, ADT 37, ADT 39, ADT 43, ADT (R) 45, ADT (R) 47, ASD 16, ASD 18, ASD 20, MDU 5, CO 47,CORH 3

Pishanam/Late Pishanam (Sep-Oct.)

ASD 18, ASD 16, CO 43, TRY 1*, ADT (R) 46, CORH 4,CO (R) 48, CO (R) 49, CO (R) 50, ADT (R) 49

b. Rainfed Lowland Rice (Dry Rice)

TKM 19, IR 50, PMK1, PMK2, MDU5, TKM11, PMK3, TPS1

c. Rainfed Lowland Rice with Irrigation (Semi Dry Rice)

ADT 36, ADT 39, TKM 10, TKM 11, TKM (R) 12, MDU 5, PMK (R) 3, Anna (R) 4, RMD (R) 1

d. Dry Seeded Upland Rice

Small batches in and around Dharmapuri district, Short duration rice varieties are used

e. Deep Water Rice

In certain pockets of Nagapattinam and Tiruvarur districts

f. Rainfed Direct Seeded (July-Aug)

MDU 5, TKM 11,PMK (R) 3, TKM (R) 12, Anna (R) 4), RMD (R) 1

7. Chemical Properties of Flooded Soils

Since rice is predominantly grown under wetland conditions, it is important to understand the unique properties of flooded soils for better management of fertilizers for this crop. When a soil is flooded, the following major chemical and electrochemical changes take place:

a) Depletion of molecular oxygen

b) Chemical reduction of soil

c) Increase in pH of acid soils and decrease in pH of calcareous and sodic soils

d) Increase in specific conductance

e) Reduction of Fe^{3+} to Fe^{2+} and Mn^{4+} to Mn^{2+}

f) Reduction of NO_3^- to NO_2^-, N_2 and N_2O

g) Reduction of SO_4^{2-} to S^{2-}

h) Increase in supply and availability of N, P, Si and Mo

i) Decrease in concentrations of water-soluble Zn and Cu

j) Generation of CO_2, methane and toxic reduction products, such as: organic acids and Hydrogen sulphide

These will have a profound influence on soil nutrient transformations and availability to rice plants.

8. Role of Nutrients for Rice Crop

Macronutrients

Functions and Mobility of N

☆ Nitrogen increases leaf size and spikelet number per panicle. N affects all parameters that contribute to yield.

☆ Leaf color, an indicator of crop N status, is closely related to the rate of leaf photosynthesis and crop production. When sufficient N is applied to the crop, the demand for other nutrients such as P and K increases.

☆ Nitrogen promotes rapid growth and increases plant height, panicle number, leaf size, spikelet number, and number of filled spikelet, which largely determine the yield capacity of a rice plant.

☆ Rice plants require N during the tillering stage to ensure a sufficient number of panicles. The critical time at active tillering for N application is typically about midway between 14 days after transplanting (DAT) or 21 days after sowing (DAS) and panicle initiation.

☆ From internodes elongation (green ring) through the beginning of head formation, nitrogen must be available in sufficient quantity to promote the maximum number of grains.

☆ Nitrogen deficiency at this time reduces the number of potential grains (florets) and limits yield potential.

☆ At panicle initiation (about 60 days before harvest of tropical rice), it is critical that the supply of N and K are sufficient to match the needs of the crop.

Figure 7.1: In the Omission Plot where N has not been Applied, Leaves are Yellowish Green.

Figure 7.2: In N-deficient Plants, Leaves are Smaller.

☆ Insufficient N at panicle initiation can result in loss of yield and profit through reduced number of spikelets per panicle.

N-deficiency Symptoms and Effects on Growth

The initial symptom of nitrogen deficiency in rice is a general light green to yellow color of the plant. It is first expressed in the older leaves because nitrogen is translocated within the plant from the older leaves to the younger ones. Prolonged nitrogen deficiency causes severe plant stunting, reduced tillering and yield reduction (Figures 7.1 and 7.2).

Phosphorus

Functions and Mobility of P

Proper phosphorus (P) nutrition is critical for producing maximum rice grain yields. Phosphorus is very important in the early vegetative growth stages.

Phosphorus promotes strong early plant growth and development of a strong root system. It is important to rice plants because it promotes tillering, root development, early flowering, and ripening.

Figure 7.3: Tillering is Reduced where P is Deficient.

Figure 7.4: Even Under Less Pronounced P Deficiency, Stems are Thin and Spindly, and Plant Development is Retarded.

Figure 7.5: Plants are Stunted, Small, and Erect Compared with Normal Plants.

Deficiency Symptoms and Effects on Growth

Rice plants that are deficient in P are stunted and dirty-dark green, and they have erect leaves, relatively few tillers, and decreased root mass (Figures 7.3–7.5).

Potassium

Functions and Mobility of K

Potassium has essential functions in plant cells and is required for the transport of the products of photosynthesis. K provides strength to plant cell walls and contributes to greater canopy photosynthesis and crop growth. Unlike N and P, K does not have a pronounced effect on tillering. K increases the number of spikelets per panicle, percentage of filled grains, and 1,000-grain weight. Insufficient K supply

Figure 7.6: Leaf Margins become Yellowish Brown.

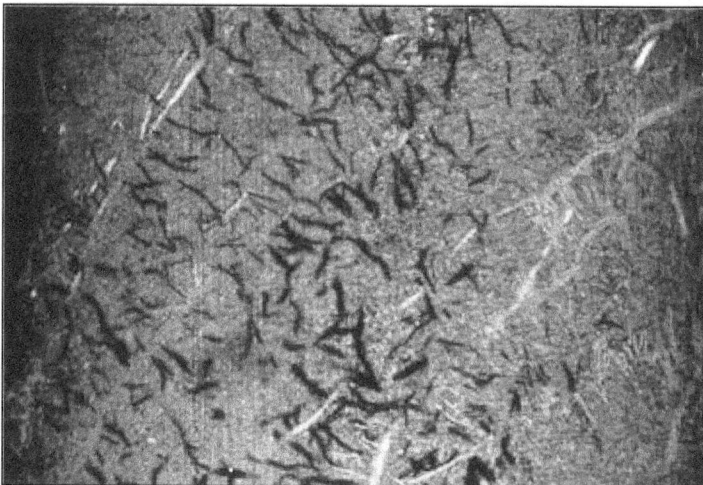

Figure 7.7: K-deficient Rice Plant Roots may be Covered with Black Iron Sulfide.

at panicle initiation can result in loss of yield and profit through reduced spikelets per panicle and reduced grain filling.

K-deficiency Symptoms and Effects on Growth

Potassium deficiency symptoms include stunted plants with little or no reduction in tillering, droopy and dark green upper leaves, yellowing of the interveinal areas of the lower leaves, starting from the leaf tips, and eventually join together across the entire leaf and turn brown on all leaves. Potassium is highly mobile in the plant, and deficiency symptoms will always occur first and be most severe on the oldest leaves. Older leaves are scavenged for the K needed by younger leaves. Dark green plants with yellowish brown leaf margins or dark brown necrotic spots first appear on the tips of older leaves (Figures 7.6–7.8).

Figure 7.8: Leaf Bronzing is also a Characteristic of K Deficiency.

Secondary Plant Nutrients

Sulfur (S)

Functions and Mobility of S

Sulfur plays an important role in the biochemistry and physiology of the rice plant, mainly in chlorophyll production, protein synthesis, and carbohydrate metabolism.

S-deficiency Symptoms and Effect on Growth

Symptoms of S deficiency are very similar to N deficiency symptoms, producing pale yellow plants which grow slowly. However, the main difference is that sulphur is immobile in the plant; therefore, the yellowing will first appear in new leaves rather than older leaves (Figures 7.9 and 7.10).

Figure 7.9: The Leaf Canopy Appears Pale Yellow Due to Yellowing of the Youngest Leaves and Plant Height and Tillering are Reduced.

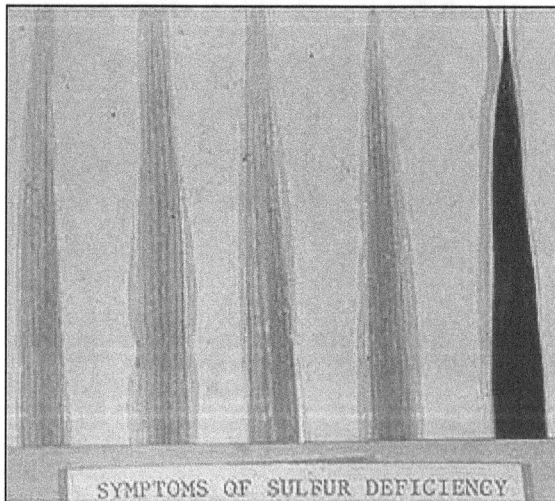

SYMPTOMS OF SULFUR DEFICIENCY

Figure 7.10: Chlorosis is more Pronounced in Young Leaves, where the Leaf Tips may Become Necrotic.

Calcium (Ca)

Functions and Mobility of Ca

Deficiency symptoms usually appear first on young leaves. Ca deficiency also results in impaired root function and may predispose the rice plant to Fe toxicity. An adequate supply of Ca increases resistance to diseases such as bacterial leaf blight or brown spot.

Calcium is important for the build-up and functioning of cell membranes and the strength of cell walls. Most calcium-related disorders of crops are caused by unfavorable growing conditions and not by inadequate supply of calcium to the roots.

Rapidly growing crops in hot windy conditions are most at risk. Deficiencies are relatively rare in irrigated rice systems, can also develop under waterlogging, soil salinity, high potassium or ammonium supply, and root disease.

Ca-deficiency Symptoms and Effects on Growth

Calcium moves in the plants' transpiration stream and is deposited mainly in the older leaves. Deficiencies, can also develop under waterlogging, soil salinity, high potassium or ammonium supply, and root disease. Chlorotic-necrotic split or rolled tips of younger leaves and stunting and death of growing points are symptoms of Ca deficiency in rice (Figure 7.11).

Figures 7.11: Symptoms Only Occur Under Severe Ca Deficiency when the Tips of the Youngest Leaves may Become Chlorotic-White.

Micronutrients

Micronutrient deficiencies typically do not occur on acid to slightly acid clay soils (pH = 5-6.5). However, silt and sandy loam soils, as well as any high-pH soils (>7.5), are subject to various micronutrients deficiencies. Soils with high available P and low organic matter are also subject to Zn deficiency.

Zinc (Zn)

Functions and Mobility of Zn

Zinc is essential for several biochemical processes in the rice plant. In plants, Zn is critical for many physiological functions, including the maintenance of structural and functional integrity of biological membranes and the facilitation of protein synthesis.

Of all micronutrients, Zn is required by the largest number of enzymes and proteins. Zn accumulates in roots but can be translocated from roots to developing plant parts. Because little retranslocation of Zn occurs within the leaf canopy, particularly in N-deficient plants, Zn-deficiency symptoms are more common on younger leaves.

Zn-deficiency Symptoms and Effects on Growth

Rice is particularly susceptible to Zn deficiency, as it grows in waterlogged soils which are conducive to zinc deficiency. Flooding the soil reduces Zn availability to the crop and increases the concentrations of soluble P and bicarbonate ions, which can exacerbate problems of Zn deficiency.

Symptoms are more severe in cold-water areas and where the flood is the deepest.

Figure 7.12: Uneven Field Growth, Plant Stunting (Foreground).

Figure 7.13: Appearance of Dusty Brown Spots on Upper Leaves.

Zinc Deficiency Symptoms

Dusty brown spots on upper leaves of stunted plants will appear 2-4 weeks after transplanting (Figures 7.12 and 7.13).

Iron (Fe)

Functions and Mobility of Fe

Iron is required for photosynthesis. Fe deficiency may inhibit K absorption. Because Fe is not mobile within rice plants, young leaves are affected first.

Fe-deficiency Symptoms and Effects on Growth

The symptoms of iron deficiency are yellowing or chlorosis of the interveinal areas of the emerging leaf. Later the entire leaf turns yellow, and finally turns white. If the deficiency is severe, the entire plant becomes chlorotic and dies. Iron deficiency can easily be mistaken for nitrogen deficiency. However, nitrogen deficiency affects the older leaves first, while iron deficiency affects the emerging

Figure 7.14: Deficiency is mainly a Problem on Upland Soils.

Figure 7.15: Symptoms Appear as Interveinal Yellowing of Emerging Leaves.

leaves first. Interveinal yellowing and chlorosis of emerging leaves is typical Fe deficiency symptom in rice (Figures 7.14 and 7.15).

Copper (Cu)

Functions and Mobility of Cu

Copper plays a key role in the following processes:

☆ N, protein, and hormone metabolism.

☆ Photosynthesis and respiration.

☆ Pollen formation and fertilization.

The mobility of Cu in rice plants depends partly on leaf N status; little retranslocation of Cu occurs in N-deficient plants. Cu-deficiency symptoms are more common on young leaves.

Cu-deficiency Symptoms and Effects on Growth

Chlorotic streaks, bluish green leaves, which become chlorotic near the tips (Figures 7.16 and 7.17).

Figure 7.16: Deficiency mainly Occurs in Organic Soils.

Boron (B)

Functions and Mobility of B

Boron is an important constituent of cell walls. B deficiency results in reduced pollen viability. Because B is not retranslocated to new growth, deficiency symptoms usually appear first on young leaves.

B-deficiency Symptoms and Effects on Growth

Boron deficiency symptoms in rice are:

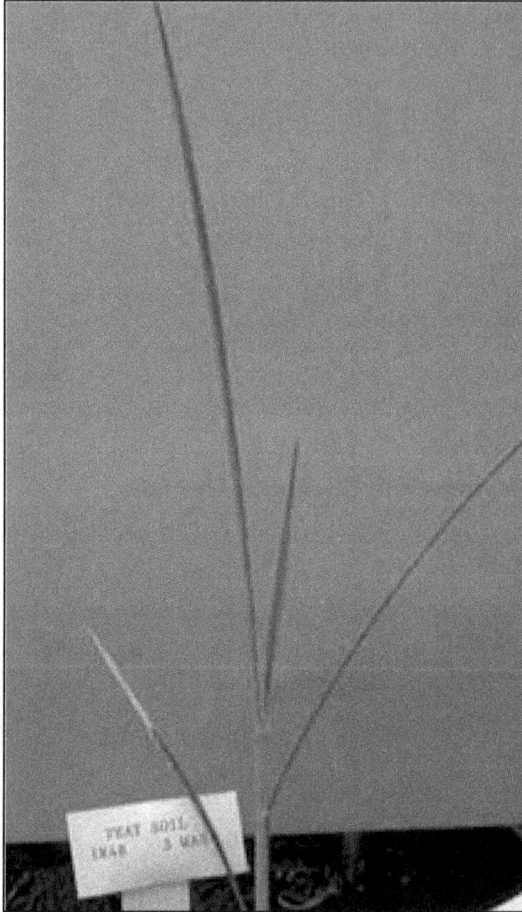

Figure 7.17: New Leaves may have a Needlelike Appearance

☆ White and rolled leaf tips of young leaves

☆ Reduction in plant height

☆ Tips of emerging leaves are white and rolled (as in Ca deficiency)

☆ Death of growing points, but new tillers continue to emerge during severe deficiency

☆ If affected by B deficiency at the panicle formation stage, plants are unable to produce panicles.

Salt Injury

Mechanism of Salinity Injury

Salinity is defined as the presence of excessive amounts of soluble salts in the soil. Na, Ca, Mg, chloride, and sulfate are the major ions involved. The effects of salinity on rice growth are

☆ Osmotic effects (water stress),

Figure 7.18: Brownish Leaf Tips are a Typical Characteristic of B Toxicity, Appearing as Marginal Chlorosis on the Tips of Older Leaves.

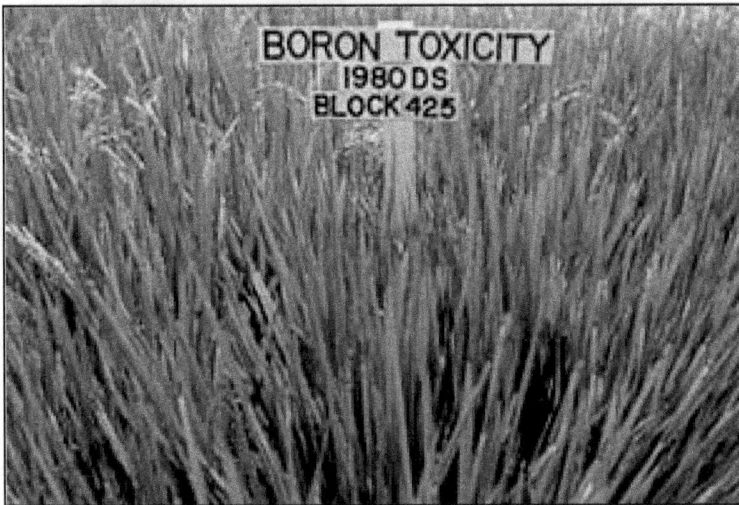

Figure 7.19: 2-4 Weeks Later, Brown Elliptical Spots Develop on the Discolored Areas.

☆ Toxic ionic effects of excess Na and Cl uptake, and

☆ A reduction in nutrient uptake (K, Ca) because of antagonistic effects.

Rice tolerates salinity during germination, is very sensitive during early growth (1-2-leaf stage), is tolerant during tillering and elongation, but becomes sensitive again at flowering.

Salinity Symptoms and Effects on Growth

White leaf tips and stunted, patchy growth in the field (Figures 7.20–7.22).

Figure 7.20: Rice Growth is Characteristically Patchy in Soils Affected by Salinity.

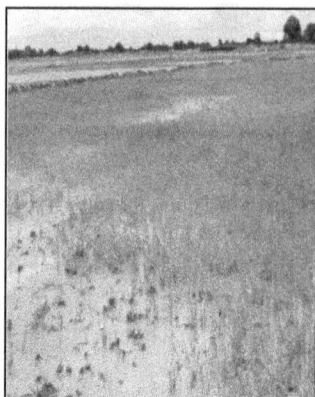

Figure 7.21: Where Saline Irrigation Water is Used, Patches of Affected Plants are found Adjacent to Water Inlets.

Figure 7.22: Plants are Stunted with White Leaf Tips.

9. Nutrient Removal by Transplanted Rice

Rice crop removes on an average of 15-20 kg N, 3-5 kg P_2O_5 and 15-25 kg K_2O apart from varying quantities of the other essential nutrients to produce one ton of grain with an equal amount of straw from the soil.

Nutrients	kg/t Grain Yield	Nutrients	kg/t Grain Yield
N	15-20	P	3-5
K	15-25	Ca	16-20
Mg	6-8	S	2-3.2
Fe	0.6-0.7	Mn	0.5-0.6
Zn	0.1-0.2	Cu	0.010-0.015
B	0.015		

IRRI (2008).

10. Nutrient Management Practices for different Rice Systems in Tamil Nadu

I. Transplanted Rice

a. Nursery Management

Seed Treatment

☆ *Pseudomonas fluorescence* (10 g/kg seed) + *Azospirillum* (1kg/ha)

☆ Phosphobacteria (1kg/ha)

Nutrient Management

☆ FYM or compost (1 ton) + 4 kg DAP/20 cents

☆ 1 kg DAP/cent 10 days prior to pulling out of seedlings

☆ For clayey soils where root snapping is a problem, 4 kg of gypsum and 1 kg of DAP/cent can be applied at 10 days after sowing

b. Main Field Management

Land Preparation

☆ Puddling and perfect field leveling

Nutrient Management

☆ Apply 500 kg of gypsum/ha at last ploughing

☆ 12.5 t of FYM/Compost or 6.25 t/ha of green leaf manure

☆ Azospirillum and Phosphobacteria each @10 packets per ha

☆ Blue Green Algae inoculation @ 10kg/ha

☆ Raise Azolla as a dual crop by inoculating 250 kg/ha at 3 to 5 DAT and incorporation during weeding for the wet season crop.

☆ Pseudomonas fluorescens (Pf 1) at 2.5 kg/ha mixed with 50 kg FYM and 25 kg of soil and broadcast the mixture uniformly before transplanting.

☆ Apply fertilizer nutrients as per STCR-IPNS recommendations for desired yield target (Appendix I) (or)

◆ N dose may be through Leaf Color Chart (LCC)*

◆ P and K may be through Site Specific Nutrient Management by Omission plot technique** (Appendix II)

☆ If the above recommendation are not able to be followed, adopt blanket recommendation as follows:

Season/Region

Short Duration Varieties (Dry Season)	N	P_2O_5	K_2O
		kg/ha	
Short duration Varieties			
a) Cauvery delta and Coimbatore tract	150	50	50
b) For other tracts	120	40	40
Medium and long duration varieties (wet season)	150	50	50
Hybrid rice	175	60	60
Low N responsive cultivars (like whit ponni)	75*	50	50

* For white Ponni, N should be applied in three splits at Active Tillering (AT), Panicle initiation (PI) and Heading stages.

☆ N and K in four equal splits *viz.*, basal, tillering, panicle initiation and heading stages

☆ P applied as basal and incorporated in soil

☆ Gypsum @ 500 kg/ha as basal

☆ Micronutrient mixture 12.5 kg/ha or 25 kg of zinc sulphate enriched in 250 kg FYM

☆ Foliar spray of 1 per cent urea + 2 per cent DAP + 1 per cent KCl at PI and 10 days later for all varieties for yield maximization

II. Wet Seeded Rice

Land Preparation

☆ On receipt of showers repeated ploughing should be carried out and after immudation puddling is to be done as per transplanting

Nutrient Management

☆ 12.5 t of FYM/Compost or 6.25 t/ha of green leaf manure

☆ Azospirillum and Phosphobacteria each @10 packets per

☆ Blue Green Algae inoculation @ 10kg/ha

☆ Raise Azolla as a dual crop by inoculating 250 kg/ha at 3 to 5 DAT and incorporation during weeding for the wet season crop.

☆ *Pseudomonas fluorescens* (Pf 1) at 2.5 kg/ha mixed with 50 kg FYM and 25 kg of soil and broadcast the mixture uniformly before transplanting.

Fertilizer Recommendation

Season/Region

Short Duration Varieties (Dry Season)	N	P_2O_5	K_2O
		——— kg/ha ———	
a) Cauvery delta and Coimbatore tract	150	50	50
b) For other tracts	120	40	40
Medium and long duration varieties (wet season)	150	50	50
Hybrid rice	175	60	60
Low N responsive cultivars (Like White Ponni)	75*	50	50

* For White Ponni, N should be applied in three splits at AT, PI and Heading stages

- ☆ N and K in four equal splits *viz.*, 21 DAS, tillering, panicle initiation and heading stages
- ☆ P applied as basal and incorporated in soil
- ☆ Gypsum @ 500 kg/ha as basal
- ☆ 25 kg of Zinc sulphate enriched in 250 kg FYM
- ☆ Foliar spray of 1 per cent urea + 12 per cent DAP + KCl at PI and 10 days later for all varieties

III. Dry Seeded Rainfed Un-puddled Lowland Rice

Land Preparation

- ☆ Dry ploughing to get fine tilt. Apply gypsum @ 1 t/ha

Seed Management

- ☆ Seed hardening with 1 per cent KCl
- ☆ Pseudomonas fluorescence 10 g/kg seed and with Azophos 1kg/ha seed or Azospirillum and Phosphobacteria @ 1 kg

Nutrient Management

- ☆ Apply 50: 25: 25 kg N: P_2O_5: K_2O/ha.
- ☆ Basal dose of 750 kg of enriched FYM (P_2O_5 @ 25 kg/ha).
- ☆ N and K in two equal splits at 20 -25 and 40 - 45 days after germination.
- ☆ Gypsum @ 500 kg/ha as basal for crusted soils
- ☆ Basal application of $FeSO_4$ at 50 kg/ha is desirable for iron deficient soil (or) apply TNAU Rainfed rice MN mixture @12.5 kg/ha as EFYM at 1: 10 ratio incubated for 30 days at friable moisture.
- ☆ Need based foliar application of 0.5 per cent $ZnSO_4$ and 1 per cent $FeSO_4$ may be taken up at tillering and PI stages.

☆ Foliar spray of 1 per cent urea + 2 per cent DAP + 1 per cent KCl at PI and 10 days later may be taken up for enhancing the rice yield if sufficient soil moisture is ensured.

Intercultural Operations

☆ Broadcast Azophos 2 kg mixed with 25 kg of FYM after rain

IV. Dry Seed Rainfed Un-puddled Lowland Rice with Supplemental Irrigation

Land Preparation

☆ Dry ploughing to get fine tilth

☆ Apply gypsum at 1 t/ha

☆ Shallow trenches (15 cm width) at an interval of 3m all along the field to facilitate draining excess water

Seed Management

☆ Seed hardening with 1 per cent KCl

☆ Pseudomonas fluorescence 10 g/kg seed and with Azophos 1kg/ha seed or Azospirillum and Phosphobacteria @ 1 kg

Nutrient Management

☆ Apply 75: 25: 37.5 kg NPK/ha

☆ Basal dose of 750 kg of enriched FYM (P_2O_5 @ 25 kg/ha).

☆ N and K in three splits at 20- 25, 40-45 and 60-65 days after germination

☆ Gypsum @ 500 kg/ha as basal.

☆ Basal application of $ZnSO_4$ at 25 kg/ha and $FeSO_4$ at 50 kg/ha as enriched FYM or apply TNAU rainfed rice MN mixture @ 12.5 kg/ha as EFYM at 1: 10 ratio incubated for 30 days at friable moisture

☆ Need based foliar application of 0.5 per cent $ZnSO_4$ and 1 per cent $FeSO_4$ may be taken up at tillering and PI stages.

☆ Foliar spray of 1 per cent urea + 2 per cent DAP + 1 per cent KCl at PI and 10 days later may be taken up.

Intercultural Operations

☆ Broadcast Azospirillum (2kg/ha) and Phosphobacteria (2 kg/ha) or Azophos mixed with 25 kg of FYM after rain. Spray Cycocel 1000 ppm under water deficit situations Azophos 2 kg mixed with 25 kg of FYM after rain.

V. Dry Seeded Irrigated Un-puddled Lowland Rice

Field Preparation

☆ Plough to get fine tilth

☆ Apply gypsum at 1 t/ha basally wherever soil crusting and soil hardening problem exist

☆ Shallow trenches (15 cm width) at an interval of 3m

Seed Treatment

☆ Seed hardening with 1 per cent KCl for 16 hours

☆ Seed treatment with Pseudomonas fluorescence 10 g/kg seed and

☆ Azophos 1 kg/ha or Azospirillum and Phosphobacteria @ 1 kg/ha

Intercultural Operations

☆ Broadcast Azospirillum (2kg/ha) and Phosphobacteria (2 kg/ha) or Azophos mixed with 25 kg of FYM after rain

Nutrient Management

☆ FYM/compost at 12.5 t/ha or 750 kg of FYM enriched with 50 kg P_2O_5

☆ Apply 75: 50: 37.5 kg N: P_2O_5: K_2O/ha

☆ N and K in three splits at 20-25, 40-45 an d 60-65 days for short duration varieties four splits for medium duration varieties at around 20-25, 40-45, 60-65 and 80-85

☆ N and K in three splits at 20-25, 40-45 and 60-65 days for short duration varieties four splits for medium duration varieties at around 20-25, 40-45, 60-65 and 80-85.

☆ N Need based foliar application of 0.5 per cent $ZnSO_4$ and 1 per cent $FeSO_4$ may be taken up at tillering and PI stages. Foliar spray of 1 per cent urea + 2 per cent DAP + 1 per cent KCl at PI and 10 days later may be taken up.

VI. SRI Method of Cultivation

Manuring in Modified Mat Nursery

Preparation of Soil Mixture

☆ Four (4) m^3 of soil mix is needed for each 100 m^2 of nursery

☆ Mix 70 per cent soil + 20 per cent well-decomposed pressmud/bio-gas slurry/FYM + 10 per cent rice hull.

☆ Incorporate 1.5 kg of powdered DAP or 2 kg 17-17-17 NPK fertilizer in the soil mixture

Filling in Soil Mixture

☆ Place a wooden frame of 0.5 m long, 1 m wide and 4 cm deep divided into 4 equal segments on the plastic sheet or banana leaves.

☆ Fill the frame almost to the top with the soil mixture

Seed Treatment with Biofertilizers

☆ Five packets (1 kg/ha) of Azospirillum and five packets (1 kg/ha) of Phosphobacteria or five packets (1 kg/ha) of Azophos.

☆ Biofertilizers are mixed with water used for soaking and kept for 4 hrs.

☆ The bacterial suspension after draining may be sprinkled in the nursery before sowing the treated seeds

Soil Application of Biofertilizers

☆ Application of Azospirillum @ 2 kg and Arbuscular mycorrhizal fungi @ 5 kg for 100 m^2 nursery area.

☆ Spraying fertilizer solution (optional): If seedling growth is slow, sprinkle 0.5 per cent urea + 0.5 per cent zinc sulphate solution at 8-10 DAS.

☆ For elite seedling production under modified mat nursery : seed fortification with 1.0 per cent KCl mixed with native soil and powdered DAP @ 2.0 kg per cent along with Pseudomonas 240 g/cent followed by drenching with 0.5 per cent urea solution on 9 DAS.

Main Field Nutrient Management

☆ As per transplanted rice, use of LCC has more advantage in N management.

☆ Green manure and farm yard manure application will enhance the growth and yield of rice in this system approach.

☆ Under sodic soils, during rotary weeding, apply Azophosmet @ 2.2 kg/ha and PPFM as foliar spray @500 ml/ha.

Other Package of Practices as Recommended to Transplanted Rice

☆ STCR based fertilizer recommendation for transplanted rice (for some selected districts) is given in the Appendix I.

Aerobic Rice

☆ Suitable variety PMK (R) 3.

☆ Optimum plant population : 50 hills per m^2 (20 x 10 cm).

☆ Green manure intercrop in aerobic rice : Daincha intercropping and incorporation at 25 DAS.

☆ Ridges and furrows.

☆ Fertilizer dose: 150: 50: 50 kg NPK/ha.

☆ N in four splits : 20 per cent at 15 DAS, 30 per cent at tillering and PI and 20 per cent at flowering or Nitrogen management at LCC value of 4.

☆ Basal application of $ZnSO_4$ at 25 kg/ha and $FeSO_4$ at 50 kg/ha is desirable wherever zinc and iron deficiency were noted (or) apply TNAU Rainfed rice MN mixture @12.5 kg/ha as EFYM at 1: 10 ratio incubated for 30 days at friable moisture.

☆ Need based foliar application of 0.5 per cent $ZnSO_4$ and 1 per cent $FeSO_4$ may be taken up at tillering and PI stages.

APPENDIX I

SOIL TEST BASED BALANCED FERTILISER RECOMMENDATION (STCR) FOR RICE

In the All India Coordinated Research Project for Soil Test Crop Response Correlation Studies in the Department of Soil Science and Agricultural chemistry, with the objective of providing scientific basis to recommend fertilizer doses for the conjoint use of chemical fertilizers and organic manures, studies are being carried out to establish the relationship between the soil test values, quantities of nutrients applied through fertilizers, organic manures and biofertilizers and crop yields adopting the 'Inductive' cum Targeted yield approach of Ramamoorthy *et al.*, (1967).The fertilizer prescription equations have been developed and test verified for some of the major rice soils of Tamil Nadu. The ready-reckoners drafted for the purpose of fertilizer recommendation for a given soil test value and yield targets are presented hereunder.

Soil type	:	River Alluvium (Noyyal series)
Season	:	*Kharif*
Yield Target	:	70 q ha^{-1}

Basic Data and Fertilizer Prescription Equations

	Basic Data				Fertilizer Adjustment Equations							
	NR (kg q^{-1})	Cs (%)	Cf (%)	Co (%)								
N	1.76	20.76	40.12	32.10	FN	=	4.39 T	−	0.52 SN	−	0.80 ON	
P	0.41	29.35	18.50	18.13	FP$_2$O$_5$	=	2.22 T	−	3.63 SP	−	0.98 OP	
K	1.50	19.83	61.50	44.65	FK$_2$O	=	2.44 T	−	0.39 SK	−	0.72 OK	

Ready Reckoner of Fertilizer Doses at Varying Soil Test Values for Specific Yield Target

Initial Soil Tests (kg ha^{-1})			Nutrients to be Added (kg ha^{-1})		
N	P	K	N	P$_2$O$_5$	K$_2$O
180	16	180	170	75	76
200	20	200	159	61	68
220	24	220	149	46	60
240	28	240	138	32	52
260	32	260	128	17	45
280	36	280	118	3	37

Soil type	:	Alluvium (Noyyal series)
Season	:	Rabi
Yield Target	:	70 q ha^{-1}

Basic Data and Fertilizer Prescription Equations

	Basic Data				Fertilizer Adjustment Equations								
	NR (kg q^{-1})	Cs (%)	Cf (%)	Co (%)									
N	1.76	20.76	40.12	32.10	FN	=	4.63	T	−	0.56	SN	−	0.90	ON
P	0.41	29.35	18.50	18.13	FP$_2$O$_5$	=	1.98	T	−	3.18	SP	−	0.99	OP
K	1.50	19.83	61.50	44.65	FK$_2$O	=	2.57	T	−	0.42	SK	−	0.67	OK

Ready Reckoner of Fertilizer Doses at Varying Soil Test Values for Specific Yield Target

Initial Soil Tests (kg ha^{-1})			Nutrients to be Added (kg ha^{-1})		
N	P	K	N	P$_2$O$_5$	K$_2$O
180	16	180	177	68	78
200	20	200	166	55	70
220	24	220	155	43	62
240	28	240	144	30	53
260	32	260	132	17	45

Soil type	:	Red – Sandy loam (Irugur series)
Season	:	Kharif
Yield Target	:	70 q ha^{-1}

Basic Data and Fertilizer Prescription Equations

	Basic Data				Fertilizer Adjustment Equations								
	NR (kg q^{-1})	Cs (%)	Cf (%)	Co (%)									
N	1.82	31.27	35.13	34.43	FN	=	5.19	T	−	0.89	SN	−	0.98	ON
P	0.84	72.92	37.11	17.81	FP$_2$O$_5$	=	2.27	T	−	4.50	SP	−	1.09	OP
K	2.14	33.55	68.81	58.00	FK$_2$O	=	3.11	T	−	0.59	SK	−	1.02	OK

Ready Reckoner of Fertilizer Doses at Varying Soil Test Values for Specific Yield Target

Initial Soil Tests (kg ha^{-1})			Nutrients to be Added (kg ha^{-1})		
N	P	K	N	P_2O_5	K_2O
150	8	150	178	100	98
170	12	170	160	82	87
190	16	190	142	64	75
210	20	210	124	46	63
230	24	230	106	28	51

Soil type	:	Red – Sandy loam (Irugur series)
Season	:	Rabi
Yield Target	:	70 q ha^{-1}

Basic Data and Fertilizer Prescription Equations

	Basic Data				Fertilizer Adjustment Equations						
	NR (kg q^{-1})	Cs (%)	Cf (%)	Co (%)							
N	1.76	20.76	40.12	32.10	FN =	4.88 T	−	0.68 SN	−	0.72 ON	
P	0.41	29.35	18.50	18.13	FP_2O_5 =	2.06 T	−	2.91 SP	−	2.27 OP	
K	1.50	19.83	61.50	44.65	FK_2O =	2.89 T	−	0.47 SK	−	0.59 OK	

Ready Reckoner of Fertilizer Doses at Varying Soil Test Values for Specific Yield Target

Initial Soil Tests (kg ha^{-1})			Nutrients to be Added (kg ha^{-1})		
N	P	K	N	P_2O_5	K_2O
150	8	150	191	100	103
170	12	170	177	89	94
190	16	190	164	77	84
210	20	210	150	65	75
230	24	230	137	54	65

Soil type	:	Black Alluvium (Adanur series)
Season	:	Rabi (Thaladi)
Yield Target	:	80 q ha^{-1}

Basic Data and Fertilizer Prescription Equations

	Basic Data			Fertilizer Adjustment Equations									
	NR (kg q^{-1})	Cs (%)	Cf (%)	Co (%)									
N	1.76	20.76	40.12	32.10	FN =	2.80	T	−	0.29	SN	−	0.89	ON
P	0.41	29.35	18.50	18.13	FP$_2$O$_5$ =	1.35	T	−	1.28	SP	−	1.78	OP
K	1.50	19.83	61.50	44.65	FK$_2$O =	2.50	T	−	0.42	SK	−	1.14	OK

Ready Reckoner of Fertilizer Doses at Varying Soil Test Values for Specific Yield Target

Initial Soil Tests (kg ha^{-1})			Nutrients to be Added (kg ha^{-1})		
N	P	K	N	P$_2$O$_5$	K$_2$O
180	16	180	144	74	99
200	20	200	138	69	91
220	24	220	132	64	83
240	28	240	126	59	74
260	32	260	121	54	66
280	36	280	115	48	57

Soil type	:	Black Alluvium (Kalathur series)
Season	:	Kharif (Kuruvai)
Yield Target	:	80 q ha^{-1}

Basic Data and Fertilizer Prescription Equations

	Basic Data			Fertilizer Adjustment Equations									
	NR (kg q^{-1})	Cs (%)	Cf (%)	Co (%)									
N	1.76	20.76	40.12	32.10	FN =	5.29	T	−	0.75	SN	−	0.89	ON
P	0.41	29.35	18.50	18.13	FP$_2$O$_5$ =	1.65	T	−	1.76	SP	−	0.78	OP
K	1.50	19.83	61.50	44.65	FK$_2$O =	2.73	T	−	0.37	SK	−	0.82	OK

Ready Reckoner of Fertilizer Doses at Varying Soil Test Values for Specific Yield Target

Initial Soil Tests (kg ha^{-1})			Nutrients to be Added (kg ha^{-1})		
N	P	K	N	P$_2$O$_5$	K$_2$O
180	16	180	182	71	97
200	20	200	167	64	90
220	24	220	152	57	82
240	28	240	137	50	75
260	32	260	122	43	68
280	36	280	107	36	60

Soil type	:	Black Alluvium (Kalathur series)
Season	:	Rabi (Thaladi)
Yield Target	:	80 q ha^{-1}

Basic Data and Fertilizer Prescription Equations

	Basic Data				Fertilizer Adjustment Equations								
	NR (kg q^{-1})	Cs (%)	Cf (%)	Co (%)									
N	1.76	20.76	40.12	32.10	FN	=	5.34 T	−	0.67 SN	−	0.73 ON		
P	0.41	29.35	18.50	18.13	FP$_2$O$_5$	=	1.90 T	−	1.86 SP	−	0.70 OP		
K	1.50	19.83	61.50	44.65	FK$_2$O	=	2.81 T	−	0.33 SK	−	0.80 OK		

Ready Reckoner of Fertilizer Doses at Varying Soil Test Values for Specific Yield Target

Initial Soil Tests (kg ha^{-1})			Nutrients to be Added (kg ha^{-1})		
N	P	K	N	P$_2$O$_5$	K$_2$O
180	16	180	199	84	110
200	20	200	186	77	103
220	24	220	173	69	96
240	28	240	159	62	90
260	32	260	146	55	83
280	36	280	132	47	77

Soil type	:	Red alluvium (Manakkarai series)
Season	:	*Kharif*
Yield Target	:	70 q ha^{-1}

Basic Data and Fertilizer Prescription Equations

	Basic Data			Fertilizer Adjustment Equations									
	NR (kg q^{-1})	*Cs* (%)	*Cf* (%)	*Co* (%)									
N	1.76	20.76	40.12	32.10	FN	=	4.25	T	−	0.60	SN	−	0.79 ON
P	0.41	29.35	18.50	18.13	FP$_2$O$_5$	=	2.71	T	−	4.39	SP	−	0.89 OP
K	1.50	19.83	61.50	44.65	FK$_2$O	=	3.83	T	−	0.60	SK	−	0.82 OK

Ready Reckoner of Fertilizer Doses at Varying Soil Test Values for Specific Yield Target

Initial Soil Tests (kg ha^{-1})			Nutrients to be Added (kg ha^{-1})		
N	*P*	*K*	*N*	*P$_2$O$_5$*	*K$_2$O*
180	12	180	147	110	122
200	16	200	135	93	110
220	20	220	123	75	98
240	24	240	111	58	86
260	28	260	99	40	74
280	32	280	87	23	62

Soil type	:	Red alluvium (Manakkarai series)
Season	:	*Rabi*
Yield Target	:	70 q ha^{-1}

Basic Data and Fertilizer Prescription Equations

	Basic Data			Fertilizer Adjustment Equations									
	NR (kg q^{-1})	*Cs* (%)	*Cf* (%)	*Co* (%)									
N	1.62	21.02	36.24	28.63	FN	=	4.47	T	−	0.58	SN	−	0.79 ON
P	0.46	27.84	17.29	6.74	FP$_2$O$_5$	=	2.66	T	−	3.68	SP	−	0.89 OP
K	2.06	27.26	50.49	34.22	FK$_2$O	=	4.08	T	−	0.65	SK	−	0.82 OK

Advances in Rice Cultivation: A Complete Guide on Rice

Ready Reckoner of Fertilizer Doses at Varying Soil Test Values for Specific Yield Target

Initial Soil Tests (kg ha^{-1})			Nutrients to be Added (kg ha^{-1})		
N	P	K	N	P_2O_5	K_2O
180	12	180	164	115	128
200	16	200	152	101	115
220	20	220	140	86	102
240	24	240	129	71	89
260	28	260	117	57	76
280	32	280	106	42	63

Note: Wherever green manure is applied @ 6.25 t ha^{-1} 38, 13 and 33 kg of N, P_2O_5 and K_2O can be reduced from the recommended fertiliser nutrient doses. For the addition of *Azospirillum* and Phosphobacteria each @ of 2 kg ha^{-1}, 15 and 10 kg of N and P_2O_5 respectively could be reduced from the recommended fertiliser nutrient doses.

APPENDIX II

SITE SPECIFIC NUTRIENT MANAGEMENT (SSNM) IN RICE

Much of the nutrients required by rice are supplied from soil and organic inputs, such as crop residues and manures. But this supply of nutrients is typically insufficient to meet the nutrient requirements for high rice yields. The use of fertilizers is consequently essential to meet the deficit between crop demand for nutrients and the supply of nutrients from soil and organic inputs.

Current fertilizer recommendations normally consist of fixed rates and timings of fertilizers for vast rice-growing areas. Such recommendations assume the deficit between nutrient demand and supply is constant among years and over vast areas. But the crop growth and crop demand for nutrients are strongly influenced by climate and crop-growing conditions, which can vary greatly among locations, seasons, and years. The supply of nutrients from soil and organic inputs can also vary greatly among rice fields. As such, even the soil test based fertilizer recommendations are not robust and precise enough to meet the variable crop needs for nutrients under different conditions.

Site-specific nutrient management (SSNM) provides an approach for 'feeding' rice with nutrients as and when needed. The rates and timing of fertilizer application are adjusted to the location- and season-specific needs of the crop and growing conditions. SSNM ensures the correct nutrients are applied at the right time and in the amount needed by the rice crop. SSNM eliminates wastage of fertilizer by preventing excessive rates of fertilization and by avoiding fertilization when the crop does not require nutrient inputs. It also ensures that N, P, and K are applied in the ratio required by the rice crop.

Principles of SSNM

The goal of SSNM is to increase profit through

* ☆ Increased yield of rice per unit of applied fertilizer
* ☆ Higher rice yields, and
* ☆ Reduced disease and insect damage.

SSNM includes two important technologies *viz.*,

1. Efficient fertilizer N management through the use of Leaf Colour Chart (LCC)
2. Omission plot technique to determine the nutrient supplying capacity of the soil and to fine tune P and K recommendations for rice.

LCC based N Management

Time of application is decided by LCC score

* ☆ Take observations from 14 DAT in transplanted rice or 21 DAS in direct seeded rice.

☆ Repeat the observations at weekly intervals up to heading

☆ Observe the leaf colour in the fully opened third leaf from the top as index leaf.

☆ Match the leaf color with the colours in the chart during morning hours (8-10 am).

☆ Take observation in 10 places.

☆ LCC critical value is 3.0 in low N response cultures like White Ponni and 4.0 in other cultivars and hybrids

☆ When 6/10 observations show less than the critical colour value, N can be applied as per the following recommendation.

Kuruvai Season

Application of 25 kg N/ha at 7 DAT followed by N @ 40 kg N/ha each time from 14 DAT as and when the leaf colour value falls below the critical value (4) for short duration varieties and hybrids.

Samba/Thaladi Season

Application of 25 kg N/ha at 7 DAT followed by N @ 30 kg N/ha each time from 14 DAT as and when the leaf colour value falls below the critical value (4) for varieties and hybrids and 3 for white ponni. For aged seedlings : Basal application of 35 kg N per ha is recommended to avoid yield loss when seedlings aged 35 - 45 days are used for transplanting and the LCC based N management can be followed from 14 DAT.

Recommendation of P and K Fertilizer Rates Based on SSNM Approach for Rice Growing Tracts of Tamil Nadu (other than Cauvery Delta)

1. Omission Plot Technique

In the SSNM approach, fertilizer P and K rates are based on the yield difference between treatments with full fertilizer use of NPK and omission plots that receives all nutrients except for the omitted (0 P, 0 K). The main objective of this technique is to determine the soil indigenous P and K supplying capacity and develop fertilizer P and K recommendations for rice in a range of soils.

2. Determination of Grain Yield

Harvest all the hills of a central 5 m² area of each treatment plot separately and properly label each harvested sample. The harvested grain will be threshed and cleaned. The grain yield is recorded at 14 per cent moisture level. The sample weight is converted to yield in t ha⁻¹.

The P and K response is determined

P response = Yield in NPK plot – Yield in P omission plot

K response = Yield in NPK plot – Yield in K omission plot

From the P and K response, P and K requirement of the field will be calculated based on the ready reckoner table as in the table below.

Recommended P_2O_5 Rates According to Target Yield and P Limited Plot

Target yield (t/ha)	4	5	6	7	8
Yield in '0' P plot	Fertilizer P_2O_5 (kg/ha)				
3	20	40	60		
4	15	25	40	60	
5	0	20	30	40	60
6	0	0	25	35	45
7	0	0	0	30	40
8	0	0	0	0	35

Example:	
Experiment	Yield (t/ka)
P_2O_5 fertilized	6
P omitted	5
Calculated P_2O_5 (kg/ha) to get an yield of 6 t/ha	30

Recommended K_2O Rates According to Target Yield and K Limited Plot

Target yield (t/ha)	4	5	6	7	8
Yield in '0' K plot	Fertilizer K_2O (kg/ha)				
3	30	60	90		
4	0	35	65	95	
5	0	20	50	80	110
6	0	0	35	65	95
7	0	0	0	50	80
8	0	0	0	0	65

Example:	
Experiment	Yield (t/ka)
K_2O fertilized	6
K omitted	5
Calculated K_2O (kg/ha) to get an yield of 6 t/ha	50

SSNM Recommendation

☆ Application of 25 kg N/ha at 7 DAT followed by 40 kg N/ha each time based on LCC – 4 from 14 DAT

☆ P and K through SSNM for Kuruvai and Thaladi season

☆ For Cauvery Old Delta – 35 kg P_2O_5 and 50 kg K_2O/ha

☆ For Cauvery New Delta – 35 kg P_2O_5 and 80 kg K_2O/ha

P and K Recommendation Based on SSNM Approach for different Tract of Tamil Nadu

Location	Calibrated SSNM P_2O_5	Fertilizer Dose (kg/ha)*K_2O
Coimbatore	30	40
Killikulam	30	50
Trichy	35	50

Contd...

Location	Calibrated SSNM P_2O_5	Fertilizer Dose (kg/ha)*K_2O
Ambasamudram	40	50
Bhavanisagar	20	25
Paiyur	25	45
Yethapur	30	45
Aruppukottai	20	30
Cuddalore	30	50

Problem Soil Management Technologies

Specific recommendations are available for the different types of problem soils of chemical nature and the details of such recommendations are furnished below :

Salt Affected Soils

☆ 25 per cent excess N and 40 kg $ZnSO_4$ ha^{-1} to all systems of rice

Acid Soils

☆ Apply lime @ 2.5 t ha^{-1} before last ploughing up to fifth crop

Fluffy Paddy Soils

☆ Compaction with 400 kg stone roller (or) oil drum with stones inside once in three years

Sodic Soils

☆ Gypsum applications at 50 per cent GR – incorporate and drain the stagnant water.

☆ Application of GLM at 5 t/ha 10 -15 DBT ; 50 per cent excess $ZnSO_4$

Saline Soils

☆ Provision of drainage

☆ Application of GLM at 5 t/ha 10 -15 DBT and 25 per cent excess N with 25 kg $ZnSO_4$ ha^{-1}

Stubble Incorporation for Thaladi in Cauvery Delta

Application of 22 kg urea/ha at the time of first puddling while incorporating the stubbles of previous kuruvai crop to compensate immobilization of N by the stubbles. This may be done at least 10 days prior to planting of subsequent crop.

Chapter 8

Water Management in Rice

B.J. Pandian

Water Technology Centre,
Tamil Nadu Agricultural University, Coimbatore, Tamil Nadu
e-mail: directorwtc@tnau.ac.in

Rice [*Oryza sativa* (L.)] is one of the most important staple food crops in the world. In Asia, more than two billion people are getting 60-70 per cent of their energy requirement from rice and its derived products. India is the second largest producer and consumer of rice in the world. In India, rice occupies an area of 44.6 million hectare with production of 90 million tonnes with productivity of 2.0 tonnes per hectare. West Bengal, Uttar Pradesh, Madhya Pradesh, Bihar, Odisha, Andhra Pradesh, Assam, Tamil Nadu, Kerala, Punjab, Maharashtra and Karnataka are major rice growing states and contribute to total 92 per cent of area and production. The productivity of rice has increased from 1984 kg per hectare in 2004-05 to 2372 kg per hectare in 2011-12.Demand for rice is growing every year and it is estimated that in 2025 AD the requirement would be 140 million tones and hence it is essential to sustain present food self-sufficiency and to meet future food requirements. Rice cultivation requires large quantity of water and for producing one kg rice, about 3000 - 5000 litres of water is required.

Crop water requirement is the water required by the plants for its survival, growth, development and to produce economic parts. This requirement is applied either naturally by precipitation or artificially by irrigation. The daily consumptive use of rice varies from 6-10 mm and total water ranges from 1100 to 1250 mm depending upon the agro climatic situation, duration of variety and characteristics of the soils Table 8.1.

Table 8.1: Range of Values of Water Requirements of rice (mm)

Item	Range of values (mm)
Land preparation and raising seedlings in nursery	40-50
Main field puddling and transplanting	200-300
Evapotranspiration in main field	500-700
Seepage and percolation	200-300
Total water requirement	790-1500

   i. The short duration rice crop requires about 1100 mm of water as follows
      Nursery: 3 per cent (33 mm) of water
      Field preparation: 16 per cent (176 mm) of water
      For entire crop growth in main field: 81 per cent (891 mm) of water

  ii. The evapotranspiration need for the rice crop will be as detailed below
      Planting to maximum tillering : 5.64 to 7.89 mm day^{-1}
      Tillering to heading: 11.52 to 12.20 mm day^{-1}
      Heading to harvest: < 4.00 mm day^{-1}
      Average loss: 6-13 mm day^{-1}

      Perfect puddling reduces 20 per cent of water requirement in the main field. The desirable water depth at different growth stages of rice is presented in Figure 8.1.

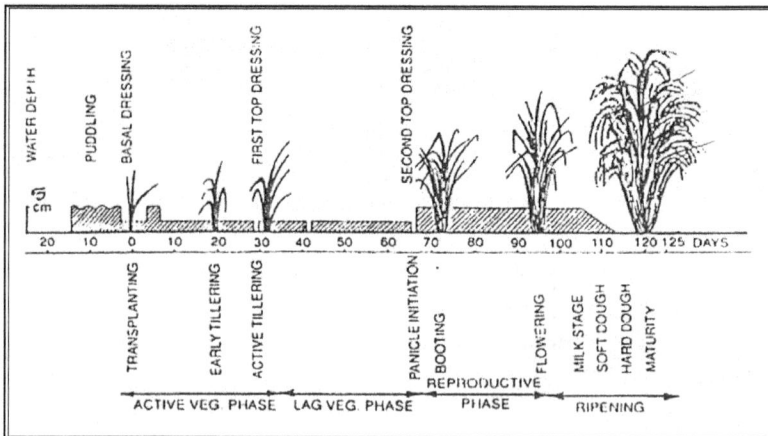

Figure 8.1: Desirable Water Depth at different Growth Stages of Rice

 iii. Comparative account of WUE of rice crop in different seasons is furnished in Table 8.2.

Table 8.2: Comparative Account of WUE of Rice Crop

Season	Duration (Days)	Water Requirement (mm)	Grain Yield (kg ha⁻¹)	WUE (kg ha⁻¹ mm⁻¹)
Kuruvai (Jun.-Sep.)	110	1263	6000	4.75
Thaladi (Sep.-Dec.)	135	918	4250	4.60
Samba (Aug.-Jan.)	155	1147	5300	4.62

SWMRI (1999).

The total water requirement was higher in Kuruvai with a higher WUE and yield.

iv. Critical stages of water requirement of rice: The most sensitive stages for water deficiency are flowering, reduction division and primordial initiation stages. Water stress at vegetative, reproductive and ripening stages visible in short and long duration varieties. In medium duration varieties, the reproductive and ripening stages are visible for water deficiency. Water deficiency at various critical stages of rice and its ill effects are presented in Table 8.3. Water deficiency at these stages reduces the crop growth and yield.

Table 8.3. Critical Stages of Water Requirement of Rice

Sl.No.	Water Deficiency at Various Stages	Ill Effects
1.	Seedling	Reduce rooting
2.	Tillering	Reduce tillering
3.	Panicle initiation to flowering	Reduce fertile grains
4.	Milk dough stage	Reduce filled grains
5.	Reduction division stage	Reduce 70 per cent yield
6.	Primordial initiation	Reduce number of grains

Rice production needs to be increased to feed the growing population whereas water for irrigation is getting scarcer. Major challenges are to (i) Save water; (ii) Increase water productivity and (iii) Produce more rice with less water.

Different Water Losses in Rice Field

Evapotranspiration

Evaporation is highest at early growth stages, when the leaf area index (LAI) is small and accounts for most of the evapotranspiration (ET) losses. In most of the tropics, the ET requirements during wet season, on an average, is 4-5 mm day⁻¹. During the dry season, it is 8-10 mm day⁻¹.

Percolation and Seepage

Percolation occurs in vertical direction and seepage by horizontal movement through levees. Percolation is largely affected by topography, soil characteristics and depth of water table. In a heavy soil with water table close to the soil surface, percolation loss could be around 1.0 mm day^{-1} as against as high as 10.0 cm day^{-1} in light soils.

Seepage, normally, flows into soil surface or streams, rivers or drains, while percolation, usually, contribute to water table.

Surface Runoff

It represents a major loss of water in lowland rice production, especially with flowing irrigation water from field to field. When the source of irrigation water is wells, there may not be any surface runoff losses. Surface runoff losses are most severe in wet season especially at times of high intensity rains.

Irrigation Management Technologies

1. Irrigation to Rice at 5 cm Depth

Irrigation rice to a depth of 5 cm when the soil reached saturation had consumed less water without affecting yield compared to continuous submergence. The saving of water worked out of 48 per cent in summer and 31 per cent in kharif. Irrigating the field when it depleted 20 per cent available soil moisture to saturation consumed the least quantity of water but the reduction in yield was 8 per cent in summer and 12 per cent in kharif.

Experiments conducted at Coimbatore and Madurai had indicated that irrigation at 5 cm to saturation recorded significantly higher grain yield with a substantial saving of water. Experiment conducted at sub centers of Water Technology Centre (WTC) showed that application of 5 cm depth of water three days after the disappearance of ponded water (DADPW) gave higher grain yield over continuous submergence besides saving about 40 per cent water under red loamy soil conditions.

Operational Research Project (ORP) on Water Management

The data obtained from ORP on water management in Karur district are presented in the following table 8.4.

Table 8.4: Quantity of Irrigation Water Used at different Reaches of Sluice Command

Sluice Reach	Quantity of Irrigation Water Applied (mm)*			
	Improved Water Management Practice		Farmers Practice	
	L_1	L_2	L_1	L_2
Head reach	1010	1040	1200	1260
Middle zone	902	955	1050	1110
Tail end	900	940	850	875

* Includes effective part of rainfall from a total rainfall of 374 mm during crop period and the water used for nursery and field preparation (250 mm).

a) Quantity of irrigation water used at different reaches of sluice command

The results revealed that the improved water management practice (irrigation to 5 cm depth three days after disappearance of previously ponded water (DADDPW) and irrigation the fields with individual field channels for irrigation and drainage found to register higher grain yield besides saving good amount of irrigation water, also facilitated equi-distribution of water to the tail end land area even under water scarcity period contrary to the farmers practice (flood irrigation).

The water saving was around 14-17 per cent in head reach and middle zone and fairly improved the distribution of water in tail end.

b) Rice grain yield (kg ha⁻¹) under improved water management and farmers practice in ORP areas during kuruvai.

Table 8.5: Effect of Improved Water Management Practices on Rice Grain Yield

Sluice Reach	Rice Grain Yield (kg ha⁻¹)			
	Improved Water Management Practice		Farmers Practice	
	L_1	L_2	L_1	L_2
Head reach	7933	6807	7222	6350
Middle zone	7545	6307	6560	5865
Tail end	7308	5930	5701	3759

SWMRI, (1998).

The yield increase was 10-15 per cent in improved irrigation practices over farmers practice. Main points to be considered for effective water management in rice are as follows.

☆ Summer ploughing

☆ Thorough puddling-levelling

☆ Trimming bunds and plugging rat and crab holes

☆ Thin sheet of water from three to seven Days After Transplanting (DAT) till establishment

☆ Irrigation at 5 cm depth one day after the disappearance of ponded water every time

☆ Creation of hand bunds for water economy avoiding seepage

☆ Weed free condition

At the time of transplanting, a shallow depth of 2 cm of water is adequate since high depth of water will lead to deep transplanting resulting in reduction in tillering. Up to seven days after transplanting, maintain 2 cm of water. During this period, establishment of seedlings take place. That is, the depletion should be recouped as and when it occurs. After the establishment stage, cyclic submergence has to be continued throughout the crop period.

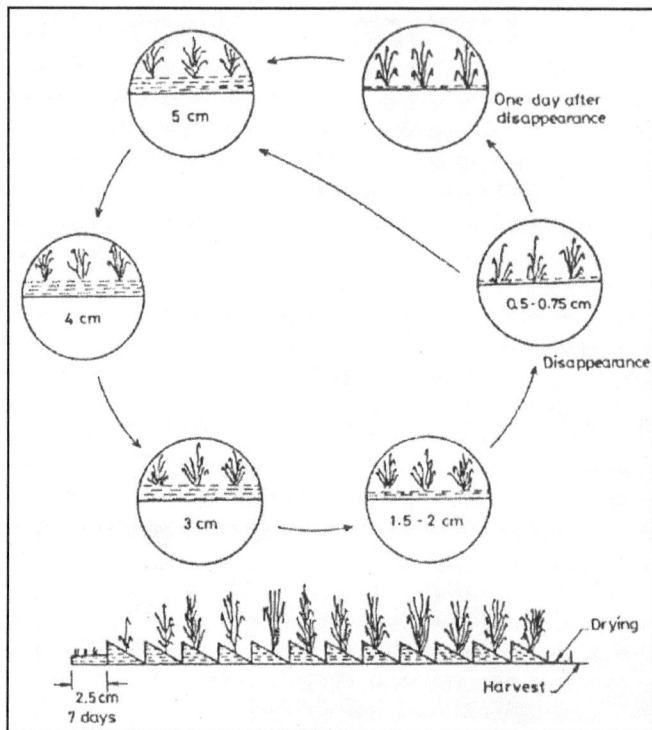

Figure 8.2: Irrigation Management for Rice Impounding 5 cm Water on Disappearance or One Day after Disapperaance.

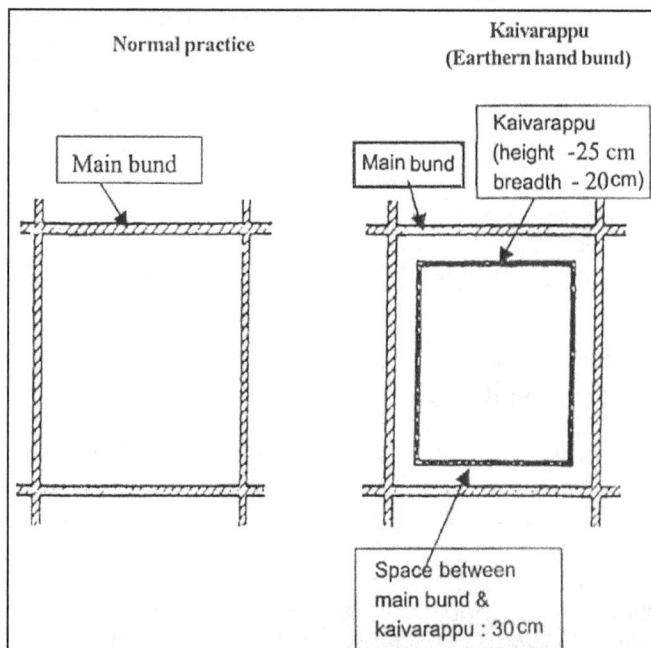

Figure 8.3: Normal Field and Field with Earthern Hand Bund.

2. Kattuthalai (or) Kaivarappu (or) Earthern Hand Bund (Figure 8.3)

The technology of forming a small hand bund of 15-20 cm at 25 to 30 cm inside the existing field bund is called Kaivarappu or Kattuthalai (Figure 8.3). This technology was tested at Erode and Salem Districts on rice under well irrigation. The total water used (irrigation + rainfall) was only 123.5 cm and 125.0 cm at Erode and Salem Districts, respectively as against 162.0 cm and 165.0 cm in control with a WUE of 4.96 kg ha^{-1} mm^{-1} and 4.84 kg ha^{-1} mm^{-1} over control which recorded 3.45 kg ha^{-1} mm^{-1} and 3.38 kg ha^{-1} mm^{-1} respectively. The per cent increase in grain yield of rice over control was 9.6 and 7.5 respectively for Erode and Salem Districts (Table 8.6).

Table 8.6: Effect of Kattuthalai on Rice Yield (Erode and Salem)

Particulars	Control	Treatment
Erode District		
Water used through irrigation (cm)	130.5	92.0
Rainfall (cm)	31.5	31.5
Total water used (cm)	162.0	123.5
Per cent water saving	–	23.8
Grain yield (kg ha^{-1})	5585	6125
Straw yield (kg ha^{-1})	5927	6316
Per cent increase in grain yield	–	9.6
WUE ((kg ha^{-1} mm^{-1})	3.45	4.96
Salem District		
Water used through irrigation (cm)	143.0	102.0
Rainfall (cm)	22.0	22.0
Total water used (cm)	165.0	124.0
Per cent water saving	–	24.8
Grain yield (kg ha^{-1})	5583	6006
Straw yield (kg ha^{-1})	6130	6215
Per cent increase in grain yield	–	7.5
WUE ((kg ha^{-1} mm^{-1})	3.38	4.84

WTC (1997).

3. Depth and Time of Irrigation for Rice Crop

The development of rice may be divided into three phases:

1. The vegetative phase, which is from germination of seed to panicle initiation, which includes germination of seed, seedling, active tillering, elongation and vegetative lag stages

2. The reproductive phase also known as flowering phase is from panicle initiation of anthesis. This includes panicle initiation, booting, heading and anthesis stages

3. The ripening phase, which is from flowering to full maturing is grain development stage.

The growth phases and stages of short, medium and long duration varieties of rice are shown in Figure 8.4.

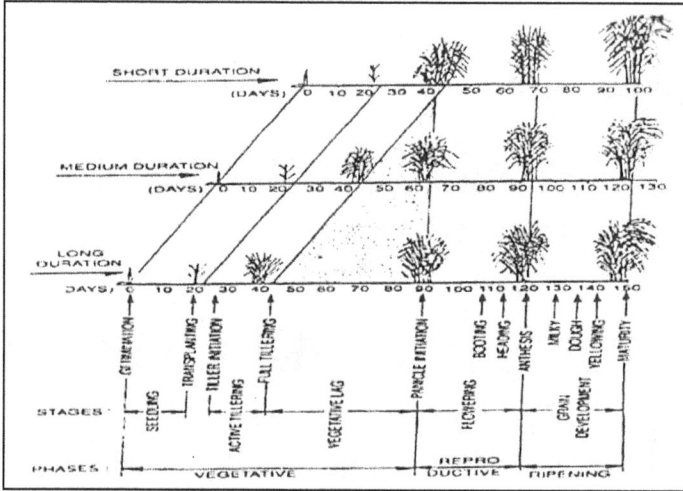

Figure 8.4: Growth Stages of Rice.

Desirable depth of submergence at various stages of rice crop is given in Table 8.7.

Table 8.7: Depth of Submergence at Various Stages of Rice

Stage of Crop Growth	Depth of Submergence (cm)
At transplanting	2
After transplanting to three days	5
Three days after transplanting upto maximum tillering	2
At maximum tillering (in fertile fields only)	Drain water for three days
Maximum tillering to panicle initiation	2
Panicle initiation to 21 days after flowering	5
Twenty one days after flowering	Withhold irrigation

Sankara Reddy and Yellamanda Reddy (1999).

Results from various experiments conducted on depth and time of irrigation for rice to answer the questions on how best the water can be saved without causing yield reduction are presented below.

If leveling of field is perfect, maintaining 2.5 cm of water throughout the crop period could result in the maximum water use efficiency (8.6 kg ha^{-1} mm^{-1}). However, weed problem will have to be circumvented by suitable techniques. An alternate method is maintaining 2.5 cm upto maximum tillering stage, draining water for two days and then maintaining 5 cm depth until 15 days prior to harvest depending upon the soil conditions.

Under a specific situation (low infiltration, evaporation and underground water replenishment) irrigation rice crop to 2.5 cm depth of water once in 10-13 days registered the maximum water production function of 15.71 kg ha^{-1} mm^{-1}.

During *thaladi,* irrigation the crop to 2.5 cm after complete disappearance of water recorded maximum grain yield of 4032 kg ha^{-1} with 507 mm of irrigation water followed by daily topping to 2.5 cm, which recorded 3972 kg ha^{-1}. The irrigation water consumption in this case was also similar (506 mm). The treatment of irrigation to 10 cm after complete disappearance has consumed 812 mm of water with a yield of 3938 kg ha^{-1} when no rains were received during the crop period.

During *kuruvai* season, irrigating 10 cm after complete infiltration and 2.5 cm daily topping were on par in terms of rice grain yield with the water consumption of 629 and 393 mm, respectively. Production potentials of the above treatments were 2.76 and 4.5 kg ha^{-1} mm^{-1} respectively. It is also seen that the production potential for the treatment of irrigation 2.5 cm after complete infiltration is 4.6 kg ha^{-1} mm^{-1}, which is the best choice in the standpoints of water economy and productivity.

4. Methylobacterium to Alleviate Water Stress

Application of Pink Pigmented Facultative Methylotroph (*Methylobacterium* sp.) as seed treatment (@ 200 g/10 kg seeds), soil application (@ 2 kg/ha) and foliar spray (@ 500 ml/ha) at panicle initiation and flag leaf stages for alleviation of water stress effects in both SRI and transplanted system of rice cultivation.This methylotroph has been tried in Cauvery delta villages of Tamil Nadu and case studies reporting better yield of crops even with irrigation once in about 20 days had been studied. Since it is available at low cost also through various centers of TNAU, the suitability of the same can be tested in study areas of Tamil Nadu to recommend its application for sustaining water stress conditions.

5. Integrated Water Management in Rice

The following are the measures in integrated water management in rice for effective water saving.

☆ Small plot size (50-60cents) with small bunds.

☆ Perfect levelling of main filed

☆ Use of 30-45 days old seedlings.

☆ Rotational water supply

☆ Azospirillum (Seed, soil and main field) treatments

☆ Use of pre-emergence herbicide

☆ Additional dose (25 per cent) of recommended N to make up volatilization loss of N

☆ Top dressing of potassium

☆ *In situ* rain water conservation by impounding more water in the field during rainy season.

☆ Strengthening field bunds

☆ Use of short duration drought resistant varieties

☆ Irrigation at critical stages

☆ Timely weeding

☆ Small channels with less depth and width

☆ Mid season drainage at maximum tillering

☆ Plugging rodent holes

☆ Spray of anti-transpirants

☆ Spray of potassium chloride

☆ Seed treatment with seed hardening chemicals

Moisture stress due to inadequate water at rooting and tillering stage causes poor root growth leading to reduction in tillering, poor stand and low yield. Critical stages of water requirement in rice are (a) primordial initiation, (b) booting, (c) heading and (d) flowering. During these stages, the irrigation interval should not exceed the stipulated time so as not to cause depletion of moisture below the saturation level. At booting stage, excess water, due to rainfall and canal supply to more than 5 cm inundation, leads to delay in heading and reduction in the growth of panicles. Adequate drainage facilities have to be provided to drain excess water. Stop irrigation 15 days ahead of harvest.

6. Direct Seeded Rice

An Integrated Water Management (IWM) technology package with component technologies was tested verified and demonstrated in direct seeded rice in the Mathur west tank command at Madurai. The results revealed the IWM technology caused improvemnt in growth characters and yield attributes and resulted in 35 per cent increased yield over traditional method with 9.2 per cent saving in irrigation water requirement. Water productivity with IWaM technology increased by Rs. 0.30 per m^3 and WUE increased by 0.6 kg ha^{-1}mm^{-1} over traditional method of irrigation management through continuous submergence of 4-5 cm.

7. Water Saving Technologies in Rice Crop Production

In Tamil Nadu, rice is the most important crop and is grown in 2.1 million hectares. Unlike other crops, rice is usually grown in flooded soils. This approximately consumes 80 per cent of the available irrigation resources of the State. Water needs of rice are two to four times more than that of the other crops of the same duration because of water loss by percolation, seepage, field preparation etc., under submerged conditions.

The daily consumptive use of rice varies from 6-10 mm and thus the total varies from 180-380 mm, transpiration 200-250 mm and percolation 200-300 mm.

A lot of yield gap is noticed between the average yield of farmers and average yield of experimental stations. In the absence of effective on-farm water management, maximization of rice yield per unit area per unit volume of water is a difficult task.

Rotational irrigation is the application of required amounts of water to fields at regular intervals. The field may often be without standing water between irrigations, but ideally the soil does not dry enough for moisture stress to develop. Rotational irrigation is often recommended to irrigate larger area with a limited water supply and to ensure better equity among water users. A major advantage of rotational irrigation is possibly more effective use of rainfall. Shallow submergence is advantageous during critical period of the crop.

8. System of Rice Intensification (SRI) Technology

1. 14- 15 days old seedlings from special and simple nursery
2. Planting single seedling per hill spaced at 20 x 20 or 22.5 x 22.5 cm
3. Use of rotary weeder at 10 days interval up to 41 -45 days after planting
4. Irrigating to 2.5 cm depth (up to panciale initiation: after small cracks develop on the soil surface; after PI: after ponded water disappear and frequent weeding upto four times with rotary hoe.

Benefits of the Methodology

☆ Drastic reduction in seed rate

☆ Synergistic root development and tillering of more than 70 tillers per hill with greater grain filling, greater pest and disease resistance and no lodging

☆ Good response both with traditional and high yielding varieties. However best response with high yielding variety

☆ No requirement of herbicide

☆ Multiple advantages of using weeder (weed incorporation, less labour for weeding, incorporation top dressed fertilizer, disturbance to the soil system, increasing tillering)

☆ More accessible to poor farmers as it is less dependant on external inputs

☆ Labour and water saving (40-50 percent)

☆ Environmental friendly - reduction in green house gases

☆ Increase in number of panicles m^{-2}, grains $panicle^{-1}$, grain and straw yield

☆ Higher net return.

Chapter 9

Weed Management in Rice

N.K. Prabhakaran and C. Nithya

Department of Agronomy,
Tamil Nadu Agricultural University,
Coimbatore – 3, Tamil Nadu
e-mail: nkpajay@yahoo.com

Pest management in general and weed management in particular has become a necessary tool in modern agriculture to meet the ever increasing needs and meet the requirements to increase the food grain production. In India, maximum cropped area (58 per cent) is under cereals, pulses and oilseeds. Among these, rice occupies a major share (24 per cent) of total cropped area. It is reported that the losses caused by weeds in cereals in India is to the tune of 9.28 million tones per year. The food loss prevented is equal to the food produced. Thus, it becomes imperative on our part to effectively manage the weeds, for increased productivity.

Rice, the most important staple food crop in India, is cultivated over an area of 40 million hectares under different situations of cultivation wherein the weed species as well as the intensity differ. Losses caused by weeds are estimated at 10 per cent of India's rice crop. Reduction in yield to the tune of 34 per cent in transplanted rice, 45 per cent in direct seeded low land rice and 67 per cent in upland rice are reported.

Successful weed control is essential for economical rice production. Weeds can reduce rice yields by competing for moisture, nutrients and light during the growing season. Weed infestations can also interfere with combine operation at harvest and significantly increase harvesting and drying costs. Weed seed contamination of rice grain lowers grain quality and may lower the cash value of the crop. As with any biological system, an effective weed management program must consider many factors that vary from crop to crop and year to year. The most important of these factors include planting date, climatic conditions, nursery preparation, seed quality, stand establishment and water management.

Predominant Weeds in Rice Ecosystem

Rice fields can be colonized by terrestrial, semi aquatic or aquatic plants. Wide variations from country to country and among the different types of rice culture have been reported; they range from more than 1800 species in South and Southeast Asia to about 30 species in Eastern Europe. In surveys in south and Southeast Asia listed 65 species in deep water rice, 194 in dry-seeded rice, 559 in transplanted rice, 558 in upland rice and 180 in wet-seeded rice. Inspite of wide variation, the number of species that constitute the major portion of the weed flora causing economic concern are only 9. They are:

Botanical Name	Common Name
Echinochloa crusgalli (L.) Beauv.	Barnyard grass
Echinochloa colonum (L.) Link.	Jungle rice
Cyperus difformis L.	Small flower umbrella plant
Cyperus rotundus L.	Purple nut sedge
Cyperus iria L.	Rice flat sedge
Eleusine indica (L.) Gaertn.	Goose grass
Fimbristylis littoralis Gaudich	Hoornah grass
Ammannia baccifera	Blistering Ammania
Marsilia quadrifoliata	The Water Fern

The major weeds that normally infest the rice crop in tropical countries are *Cyperus* sp., *Echinochloa crusgalli, E. colonum* etc. In Tamil Nadu the grass species *E. crusgalli, E. colonum, Panicaum* sp., the sedges *Cyperus iria, C. difformis, Fimbristylis miliacea, Scirpus* sp. and the broad leaved and aquatic weeds *Eclipta alba, Astracantha longifolia, Ammannia baccifera, Marsilia quadrifoliata, Monochoria vaginalis, Ludwigia parviflora* are the common species. However, under semi dry conditions where the crop is raised as rainfed upland crop in the early stages and later under flooded condition through irrigation from tanks the weed flora slightly differs. *Cyperus rotundus* is the most predominant weed in this situation.

Impact of Weeds on Rice

With the introduction of dwarf indica rice types with semi erect leaves the canopy does not develop as in tall indica types. The competition of weeds has therefore assumed greater significance. Added to this, heavy manuring provides favourable environment not only for the crop to grow but the weeds as well. Rice is grown under different situations in different parts of the country. The critical period for weed free condition for higher productivity is reported to be 30 - 35 days in transplanted rice whereas under direct seeded low land and upland conditions the weed free period ranges anywhere from 40 to 60 days. *Echinochloa crusgalli* competes with rice crop from sowing or planting till the harvest of the crop whereas the sedge weed, *Cyperus difformis* competes with the crop only in the early period because of its shorter life cycle.

In direct seeded conditions weeds emerge simultaneously rather earlier than rice seeds emergence and compete for soil moisture and nutrients. Weed growth in direct-seeded rice is severe and is one of the serious limiting factors in realising the yield potential of direct-seeded rice. The risk of crop yield loss due to competition from weeds by direct seeding methods is higher than transplanted rice because of the absence of the size difference between the crop and weeds and the suppressive effect of standing water on weed growth at crop establishment. Due to their better competitive ability the weeds grow more rapidly than rice seedlings. Under transplanted conditions, flooding and puddling control weeds before seedlings are planted. By the time new flesh of weeds emerge 2 to 3 weeks after transplanting, the transplanted rice seedlings establish well and assumes greater competitive ability.

Management Options

Cultural Methods

☆ Cattle grazing during the fallow period eliminate seeding by the annual weeds.

☆ Repeated ploughing with the receipt of summer showers is an useful practice of control weeds especially in direct seeded rice.

☆ In lighter soils stale seed bed preparation of stirring or shallow preparation

☆ Tillage practices should be timed (e.g., 10-14 days between passes) such that weeds have time to germinate in between tillage operations and thus killed by the succeeding operation

☆ Rotate crops and weed control practices to decrease weed build-up

☆ Before sowing through harrowing helps in reducing weed populations

☆ Method of planting

☆ Transplanting gives the plant a competitive advantage against weeds

☆ Direct seeding in rows to facilitate hand or mechanical weeding

☆ Higher populations of rice increase shading and reduce weed growth, but may increase crop lodging

☆ Manage water combined with good leveling, maintenance of a water layer in the paddy reduces the pressure of many weeds

☆ Stop weeds from setting seed during the crop cycle ("1 year seeds, 7 year weeds")

☆ Make sure weeds do not set seed during or after harvest

☆ Clean irrigation systems prevent weeds from growing along bunds and irrigation canals and thus passing of weeds along the irrigation system to adjacent field

☆ Manage the fallow to stop weeds from setting seed during fallow periods

Mechanical Methods

Weeding by machine involves the use of hand-pushed or powered weeders and is feasible only where rice is planted in straight rows. Conventional single-row rotary weeders are difficult to use because they must be moved back and forth. The IRRI-developed single and double row cono weeder can uproot and bury the weeds and are faster. Mechanical weeding should be supplemented by hand pulling the weeds that are close to rice plants.

Chemical Methods

Chemical weed management in rice deals with site-specific evaluation of herbicide alone or in combination with other weed control practices. Herbicide use is one of the most labour saving innovations that have been introduced in rice farming. For successful and economical use, it is important to understand the types of herbicides, herbicide selectivity, time and method of application techniques as well as their limitations. The herbicides commonly used for the control of rice weeds are: Pre-emergence - anilofos, butachlor, fenoxaprop, oxadiazon, oxyflourfen, pendimethalin, pretilachlor, thiobencarb; post-emergence - bentazon, 2,4-D and propanil.

Contact herbicides like paraquat at 0.5 to 0.75 kg/ha applied a week to 10 days prior to flooding and puddling desiccates the standing weeds and helps in minimal tillage. Several chemicals have been tried for weed control in transplanted rice. Alachlor, anilofos, butachlor, thiobencarb, fluchloralin, molinate, oxadiazon, oxyfluorfen, pendimathalin, are the important pre-emergence weed killers.

Propanil is established as the best chemical for post-emergence use in rice and effective on annual grasses including *E. crusgalli, E. colonum, Panicum* sp.etc. Split application of propanil 2nd and 4th week after transplanting seems to have added advantage as weeds which germinate in flushes are controlled. 2, 4-D is effective against established broad leaved weeds. It is useful as post-emergence herbicide. When used 4 weeks after transplanting it has an added advantage of reducing the emergence of unproductive tillers.

Combination of thiobencarb and 2, 4-D EE at post-emergence has activity on a wider spectrum of weeds than either of them applied alone. Similarly PE fluchloralin at 0.4 kg/ha in combination with 0.4 kg 2, 4-D EE applied as post-emergence has effective weed control on broad leaved and grass weed control.

The results from All India co-ordinated Rice Improvement project (AICRIP) indicate the usefulness of thiobencarb (1.0 kg/ha), oxadiazon (0.75 kg/ha) and butachlor (1.0 kg/ha). The yield ranged from 4.02 to 4.30 t/ha at Coimbatore under directs sown puddled conditions. At Aduthurai under transplanted conditions thiobencarb 1.0 kg/ha recorded the higher grain yield however the other chemicals as well were comparable.

The pre-emergence herbicides are effective for about 3 weeks. Stray weeds that establish before application of pre-emergence herbicides are not controlled. A hand weeding 3 weeks after herbicide application is beneficial in effective control of weeds. Thus pre-emergence application of herbicide followed by a hand weeding is a good management practice.

The efficient nitrogen management helps in keeping weeds under check. Several reports are available on reduced incidence of weeds when N is deep placed (5-10 cm) as urea super granule. Reducing or avoiding basal application of nitrogen, when seedlings from well-manured nurseries are used, keeps the weeds under check. Similar information on the reduced incidence of weed with the use of modified urea forms like slow release fertilizers and nitrification inhibitors are not lacking.

An integrated approach of summer ploughing, effective land preparation and leveling the field, good water management, control of the weeds at the nursery level, the use of pre-emergence herbicides at reduced rate and weeding 3 to 4 weeks later, reducing or avoiding basal application of nitrogen will help in effective weed management in transplanted rice.

Under non-puddled upland conditions application of pendimethalin (0.75 kg/ha) 8 days after sowing or thiobencarb + propanil (1.00 and 2.00 kg/ha) 16 days after seeding are highly effective. Hand weeding has to be combined to have effective weed control.

Nursery

Apply any one of the pre-emergence herbicides *viz.*, butachlor 50 per cent EC 1kg/ha, thiobencarb 50 per cent EC 1 kg/ha, pendimethalin 30 per cent EC 0.75 kg/ha, anilofos 50 per cent EC 0.75 kg/ha on 8th day after sowing to control weeds in the lowland nursery. Keep a thin film of water and allow it to disappear. Avoid drainage of water. This will control germinating weeds.

Transplanted Rice

☆ Conventional tillage of one dry ploughing and two passes of cage wheel puddling combined with pre-emergence application of butachlor at 1.25 kg/ha increased the productivity and profitability of rice-rice cropping with better weed control efficiency under lowland situation.

☆ Though conventional tillage with butachlor adjusted as best treatment, minimum tillage with glyphosate and butachlor application and conservation tillage with glyphosate and butachlor preformed equally for most of the parameters as well as yields and economics.

☆ Crop growth and yield were enhanced by butachlor 0.75 + 2,4-DEE 0.40 kg/ha with 100 per cent inorganic nitrogen.

☆ Pre emergence herbicide butachlor 1.0 kg/ha followed by weeding using finger type single row and double row rotary weeders resulted in higher grain yield and net profit.

☆ Butachlor 1.0 kg/ha for nursery and anoilofos + ethoxysulfuron (Tank mix) 0.25+0.01 kg/ha for main field recorded lower weed density and dry weight with higher weed control efficiency (Table 9.1).

Table 9.1: Effect of Integration of Weed Management in Rice Nursery and Main Field on Weed Control, Yield and Net Income in Transplanted Rice

Treatment		Weed Density (No./m^{-2})	Weed Dry Weight (kg/ha)	WCE (per cent)	Grain Yield (kg/ha)	Net Income (Rs/ha)
Nursery Field	Main Field					
Butachlor 1.0 kg/ha	Butachlor 1.0 kg/ha	1.72(52)	212	59.5	5688	15477
Butachlor 1.0 kg/ha	Anilofos+Ethoxy sulfuron @ 0.25 + 0.01 kg/ha	1.72(53)	193	63.2	5604	14816
Unweeded	HW once (25 DAT)	2.06(112)	298	43.1	5295	13780
Unweeded	HW twice (20 and 40 DAT)	1.44(27)	123	76.5	5614	14456

Figures in parenthesis are original values.

☆ Pre-emergence application of butachlor 0.75 kg/ha + bensulfuron methyl 50 g/ha on 3 DAT + HW on 45 DAT recorded lower weed density and weed dry weight with higher grain yield and net returns (Table 9.2).

Table 9.2: Effect of Weed Management Practices on Weed Characters, Grain Gield and Net Return of Transplanted Rice (Mean of 5 locations)

Treatment	Weed Density (No./m)		Weed Dry Weight (g/m)		Grain Yield (kg/ha)	Net Return (Rs./ha)
	25 DAT	45 DAT	25 DAT	45 DAT		
Hand weeding (HW) on 25 and 45 DAT	17	15	5.44	4.64	5650	8050
Butachlor 1 kg/ha on 3 DAT - HW on 45 DAT	4	7	2.50	3.20	5880	11953
Butachlor 0.75 kg/ha + bensulfuon methyl 50 g/ha on 3 DAT + HW on 45 DAT	3	5	1.2	1.50	6240	14478

☆ System of rice intensification and pre-emergence application of pyrazosulfuron-ethyl 30 g/ha at 3 DAT/8 DAS + weeding with finger type double row rotary weeder at 40 DAT/S recorded higher weed control efficiency and grain yield with better economic returns during both the seasons of study (Table 9.3).

Table 9.3: Effect of different Establishment Techniques and Weed Control Practices on Weed Control Efficiency (WCE) and Grain Yield (kg/ha) in Rice (*Rabi* and *Kharif*, 2009)

Treatment	WCE (per cent) at 60 DAT		Grain Yield (kg/ha)	
Establishment Method	Rabi, 2008-09	Kharif, 2009	Rabi, 2008-09	Kharif, 2009
Pyrazosulfuron ethyl at 30 g/ha + rotary weeding	76.84	79.87	4645	4718
Cono weeder	69.48	60.80	4947	4555
HW (Twice)	75.76	59.98	4803	4552

Wet Seeded Rice

☆ Productivity and economic returns of wet seeded rice with dual cropping of danicha could be maximized by the pre-emergence application of pretilachlor + safner at 0.45 kg/ha followed by one cono weeder weeding in between rows and manual weeding within the rice rows on 35 DAS in lowland conditions.

☆ Higher productivity of wet direct seeded (drum seeded) rice could be achieved by integrating intercropping of daincha and pre-emergence application of butachlor at 1.0 kg or pretilachlor + safener at 0.45 kg/ha on 4 DAS followed by one hand weeding on 35 DAS.

☆ IWM method of combining cultural method (drum seeding + daincha and chemical method (pretilachlor(s) at 450 g/ha) proved to be effective and economical method of weed control in wet seeded rice.

☆ Pre-emergence pretilachlor 0.45 kg/ha on 3 DAS + roto cylindrical weeder weeding on 45 DAS in wet seeded rice resulted in excellent control of weeds like *Echinochloa crusgalli, Panicum repens, Eclipta alba* and *Monochoria vaginalis* and higher grain yield, net monetary return and B: C ratio.

☆ Stale bed preparation by pre-puddling minimum tillage with glyphosate combine with post-plant pre emergence butachlor 1.25 kg/ha resulted in increased rice grain yield, net income and B: C ratio in rice-rice cropping.

☆ Lower density and dry weight of weeds were recorded in pre-emergence application of butachlor 1.5 kg/ha + one hand weeding (Table 9.4). The weed control efficiency, yield parameters, yield and economic returns were higher in the same treatment.

Table 9.4: Effect of Weed Control Methods on Weed Dry Weight (g/m) and Weed Control Efficiency (per cent) in Direct Seeded Rice

Treatment	Total Weed Dry Weight (g/m)		WCE (per cent)	
	30 DAS	60 DAS	30 DAS	60 DAS
PE Pretilachlor-S 0.5 kg/ha	1.19 (13.7)	1.94 (85.1)	80.9	72.7
PE Butachlor 1.5 kg/ha + one hand weeding	1.04 (8.9)	1.84 (66.4)	87.7	78.7
Hand weeding twice (Weed free)	1.08 (10.0)	1.86 (70.2)	86.1	77.5

Figures in parenthesis are original values.

☆ Pre-emergence application of pyrazosulfuron at 25 g/ha recorded lesser weed density, dry weight and higher grain yield in direct wet seeded rice (Table 9.5).

Table 9.5: Effect of Weed Management Treatments on Weed Density, Total Weed Dry Weight and WCE at 25 DAS in Direct Wet Seeded Rice

Treatment	Weed Density (No./m)	Weed Dry Weight (g/m)	WCE per cent	Grain Yield (kg/ha)
Pyrazosulfuon 25 gha + HW 45DAS	1.79 (59.00)	0.76 (3.74)	96.82	4333
Pretilacholr-s 750 g/ha + HW 45DAS	1.68 (45.67)	0.93 (6.52)	94.45	4000
HW 20 and 40 DAS	2.05 (110.33)	1.06 (9.48)	75.78	4167

Figures in parenthesis are original values.

☆ Weed control efficiency, yield attributes and yield were higher with the application of grass herbicide Metamifop 10 EC at 100 and 125 g/ha at 2 - 3 leaf stages and Metamifop 10 EC at 125 g/ha at 5 - 6 leaf stages when compared to other treatments in direct seeded rice (Table 9.6).

Table 9.6: Effect of Metamifop 10 EC on Weed Control Efficiency (per cent) and Grain Yield in Direct Seeded Rice

Treatment	Kharif, 2008-09		Kharif, 2009-10	
	WCE (per cent)	Yield (kg/ha)	WCE (per cent)	Yield (kg/ha)
T_1- Metamifop 10 EC at 100 g/ha	96.3	4124	96.2	4280
T_2- Metamifop 10 EC at 125 g/ha	96.8	4328	97.2	4401
T_3- Metamifop 10 EC at 100 g/ha	86.7	3860	86.4	3560
T_4- Metamifop 10 EC at 125 g/ha	96.5	4255	96.4	4362
T_5- Cyhalofop butyl 10EC at 100 g/ha @ 15 DAS	86.2	3728	87.0	3667

T_1 and T_2 - Application at 2-3 leaf stage of grass weeds.

T_3 and T_4 - Application at 5 - 6 leaf stage of grass weeds.

☆ Pre emergence Pretilachlor (S) 0.45 kg/ha on 3 DAS fb azimsulfuron 50 DF 35 g/ha on 20 DAS + hand weeding on 45 DAS for broad spectrum weed control and higher grain yield and economic returns in both irrigated and rainfed direct seeded rice.

Semi Dry Rice: Dual Cropping

☆ Recommended practice of normal sowing at 20 cm spacing + pendimethalin 1 kg/ha + one HW on 30 DAS or with two HW on 30 and 60 DAS produced comparable yield.

☆ Paired row sowing (15/30 cm) of rice + daincha under weed free condition gave higher net income and B: C ratio. From the results, it is concluded that dual cropping of green manure is efficient in weed management and increasing the productivity of semidry rice.

Table 9.7: Dual Cropping of Green Manure on Weed Control Efficiency (per cent) and Grain Yield (kg/ha) of Semidry Rice

Treatment	WCE (per cent)	Grain Yield (kg/ha)
PRS of rice + daincha (incorp. with 2,4-D)	17.9	3106
PRS of rice- weed free	44.4	3345
PRS of rice + daincha (manual incorp.)	51.6	3528
Normal sowing of rice- herbicide +1 HW on 30 DAS	35.0	3260

☆ Broad spectrum weed control could be achieved in paired row sowing (PRS) of semidry rice by dual cropping of *Sesbania* and incorporating at 30

DAS along with manual weeding. This treatment improved the growth, yield and economics when compared to the presently recommended methods of either manual weeding twice or pre-emergence application of herbicide followed by hand weeding (Table 9.7).

Drum Seeded Rice

☆ Drum seeded rice intercropped with green manure (dhaincha) reduce the weed density and growth while promoted the growth, yield and economic indices of drum seeded lowland rice.

☆ Combination of drum seeded rice intercropped with green manure dhaincha along with pre-emergence herbicide application of pretilachlor (30.7 EC) @ 0.45 kg/ha + safener on 5 DAS was the best weed control method on the basis of better weed control, crop yield and economic indices in drum seeded rice (Table 9.8).

Table 9.8: Effect of Weed Control in Drum Seeded Rice Under Low Land Ecosystem on Weed Control Efficiency (per cent) and Grain Yield

Treatment	WCE (per cent)			Grain Yield (kg/ha)
	40 DAS	60 DAS	Maturity	
Cyhalofop-butyl	60.9	66.3	59.7	4629
Pretil + Safener	67.7	82.2	72.1	5155
Hand weeding (20 and 40 DAS)	59.3	66.6	52.6	4695

Data not statistically analysed.

Aerobic Rice

☆ Higher crop yield and B: C ratio were obtained with PE pendimethalin 1.0 kg ha^{-1} along with single tyne sweep weeding which was comparable with PE along with hand weeding.

Chapter 10

Insect Pest Management in Rice

R.P. Soundararajan, S. Suresh and S. Kuttalam

Department of Agricultural Entomology,
Tamil Nadu Agricultural University
Coimbatore – 641 003, Tamil Nadu
e-mail: soundarrajan.rp@tnau.ac.in

Rice is an important staple food for more than 65 per cent of the people in India and cultivated all over the country. Though self sufficiency is achieved in the production of rice in our country, the growth of rate of rice production does not match the growth rate of population increase. The future demand of 22-25 million tones over the current 105 million tones should attained by improved cultivation methods. Insect pests are the one of the major constraint in rice production pose serious challenges in improving the productivity and achieving sustainability. Acquiring knowledge on various insects affecting rice and thereby adopting proper management methods will avoid yield looses by these pests. There are more than 100 insect species have been reported in rice and causing damage. However, among them 8-10 insect species are more common and causing serious damage in rice crop at Tamil Nadu. Integrated pest management denotes apart from insecticides use all pest management tactics should be included to save the ecosystem and beneficial insects.

The major insect pests affecting rice crop and their management tactics are presented below.

Green Leafhopper

The insect pest occurs in both nursery and transplanted main field. Affected plants become pale yellow in colour and get stunted in growth. If the plants are tapped large number of leafhoppers may be seen jumping to water. Both nymphs and adults suck the plant sap from the leaf and leaf sheath. Mild infestation reduces the vigour of the plant and the number of reproductive tillers. Heavy infestation causes withering and complete drying of the crop. It transmits plant diseases such as dwarf, transitory yellowing, yellow dwarf and rice tungro virus of which tungro virus causes serious problem in certain areas. The nymphs are soft bodied, yellow white in colour and later turn to green. Adults are bright green with variable black markings, wedge shaped with a characteristic diagonal movement. Grass and other weeds in the rice field serves as alternate host for this insects.

ETL

At vegetative stage or nursery: 60 Nos/25 sweepings or 5 Nos/hill, At flowering stage: 10 Nos/hill, In tungro endemic areas 2 Nos/hill.

Management

☆ Cultivate resistant varieties IR 50, PY 3, CR 1009 and white ponni

☆ The disease affected plants should be removed from the field periodically

☆ Installing light trap in the field during early night hours to attract the insects

☆ Spraying of Spraying of profenophos 50 per cent EC 1000ml or imidacloprid 17.8 per cent SL 100 ml or thiamethoxam 25 per cent WG 100 g for one hectare area

Leaf Thrips

The pest causes damage in nursery and young plants in the field. The seedlings in the affected nurseries appear as pale yellow colour with brown tips. On passing the wet palm over the top of the seedlings a large number of black adults and yellowish nymphs may be seen striking to the palm. Both the adults and nymphs lacerate the tender leaves and suck the plant sap. As a result fine yellowish lines or silvery streaks are seen on the leaves. Later, the leaves curl longitudinally and begin to dry from the tip downwards. In severe cases, the entire nursery may dry up and fail to produce seedling. Sometimes transplanted crop is also affected in the early stages.

Management

☆ Spraying of 5 per cent neem seed kernel extract in the nursery

☆ Apply 50 ml of phosphomidon 40 per cent SL or 40 ml of monocrotophos 36 per cent SL for 20 cent area

Swarming Caterpillar

It is a serious pest in nurseries and if the population is more the seedlings are completely eaten away by the caterpillars' overnight. Caterpillars march in large numbers in the evening hours and feed on the leaves of paddy seedlings till the morning and hide during daytime. They feed gregariously and after feeding the plants in one field march onto the next field. Under severe infestation crop gives the appearance of grazed plants. Attacked plants are reduced to stumps. Nurseries situated in ill-drained marshy areas attacked are earlier than dry ground. Damage is severe during July to September. The larvae is light green with yellowish white lateral and dorsal stripes in the early stages and later become dark brown or grayish green in colour.

Management

☆ Flood the nursery to expose the hiding larvae to the surface and thus they are picked up by birds.

☆ Apply Kerosene to water to suffocate and kill the larvae.

☆ Allow the ducks into the filed to feed on the larvae

☆ Drain the water from the nursery and spray chlorpyriphos 20 per cent EC 80 ml for 20 cents during late evening.

Whorl Maggot

Whorl maggots cause damage in the main field after transplanting. Damaged leaves became distorted and broke-off in the wind. Infested plants are stunted. It cause damage to the boot leaf and developing panicles, which resulted in producing only partially filled/half filled grains. Small puncture appear in the middle of the flag leaf and its margin get discolored. The maggots are found to feed on the unopened leaves and to nibble the inner margins of the leaves, which showed conspicuous feeding lesions in the lines. When leaves emerge from the whorl damage can be seen as pinholes in the leaves and white and yellowish lesions on the leaf edge.

ETL

25 per cent damaged leaves.

Management

☆ Spray chlorpyriphos 20 per cent EC @ 1250 ml or fipronil 5 per cent SC 1500 ml or phosalone 35 per cent EC 1400 ml for one hectare

Brown Planthopper

This is an important pest of rice crop in the main field. The incidence can occur from early vegetative stage to maturity stage. The symptoms will not be visible from outside in the early stages, but if we enter the field and tap the plants large number insects can be seen. They are visible only when the damage has been severe, the plants present a burnt up appearance, hopper burn, in circular patches. Both the nymphs and adults remain at the ground level and suck the plant sap. At early infestation, circular yellow patches appear which soon turn brownish due to the drying up of the plants. The patches of infestation then may spread out and cover the entire field. The grain setting is also affected to a great extent. It also acts as vector of the virus diseases like grassy stunt, wilted stunt and ragged stunt. Outbreak of this insect pest is favoured by growing susceptible varieties, closer spacing, heavy application of nitrogen coupled with cloudy weather and continuous drizzling

ETL

1 No./tiller; 2 Nos when spider is present at 1 No./hill.

Management

☆ Use resistant varieties like ADT 37, PY3, CO 42, PTB 33 and PTB 21.

☆ Avoid close planting and provide 30 cm rogue spacing at every 2.5 m to 3.0 m reduce the pest incidence.

☆ Avoid use of excessive nitrogenous fertilizers.

☆ Control irrigation by intermittent draining.

☆ Set up light traps to monitor pest population and to control.

☆ Avoid use of insecticides causing resurgence and synthetic pyrethroids

☆ Encourage natural enemies such as spiders, mirid bugs, lady bird beetles, dragon flies in the field by avoiding broad spectrum insecticides

☆ Spray 3 per cent neem oil @ 15 lit/ha or 0.03 per cent azadirachtin @ 1000 ml/ha

☆ Spray buprofezin 25 per cent SC 800 ml or dichlorvos 76 per cent WSC 500 ml or fipronil 5 per cent SC 1000 ml or imidacloprid 17.8 per cent SL 100 ml/ha.

☆ Drain the water before the use of insecticides and direct the spray towards the base of the plants for better control of the planthopper insects

Rice Yellow Stem Borer

The incidence of this insect pest is mild in the season June to September, but later on gets intensified from October to January and February. It causes damage both at vegetative and reproductive stage. In the vegetative stage, 'dead hearts' seen in the affected tillers and in the reproductive stage, 'white ear' may be seen. The caterpillar enters the stem and feeds on the growing shoot. As a result the central shoot dries up and produces the characteristic dead heart. The tillers may get affected at different stages. When they are affected at the time of flowering the ear heads become chaffy and are known as white ear. The larva feed inside the stem and pupates there itself. The adult moth are bright yellowish brown with a black spot at the centre of the forewing and a tuft of yellow hairs at the anal region. The activity of adults is indicator for the incidence of the insect pest.

ETL

During vegetative stage - 2 egg mass per square meter plants or 10 per cent dead heart; at reproductive stage – 2 per cent white ear symptoms

Management

☆ Grow resistant varieties like TKM 6, IR 20 and IR 26.

☆ Clip the tip of seedlings before transplanting to eliminate egg masses

☆ Don't use central shoot dried seedling during transplanting

☆ Avoid close planting and continuous water stagnation at early stages.

☆ Collect and destroy the egg masses, pull out and destroy the affected tillers.

☆ Set up light traps to attract and kill the moths.

☆ Release of egg parasitoid, *Trichogramma japonicum* thrice @ 5 CC/ha/ release

☆ Apply botanical pesticide 0.03 per cent azadirachtin @ 1000 ml/ha

☆ Spray chlorpyriphos 20 EC 1250 ml/ha or chlorantraniliprole 18.5 per cent SC 150 ml/ha or thiocloprid 21.7 per cent SC 500 ml/ha

Leaffolder

This is an important pest on rice and incidence starts during vegetative stage. The leaves folded longitudinally or transversely with silk and scrapped patches in such places. Larvae remain inside the fold and scrapping off green portion of the leaves leaving white patches. The plants become dry and in severe infested fields the crops started drying. The larva is yellowish green in colour and translucent. The adults are small yellow coloured moths with dark wavy lines on both pairs of wings.

ETL

At vegetative stage 10 per cent damaged leaves, during flowering stage 5 per cent damaged leaves (flag leaf)

Management

☆ Keep the bunds clean by trimming them and remove the grassy weeds.

☆ Avoid use of excessive nitrogenous fertilizers

☆ Set up light traps during early night hours to attack and kill the months.

☆ Release *Trichogramma chilonis* thrice @ 5 CC/ha

☆ Spray 0.03 per cent azadirachtin 1000ml/ha during early stage of infestation

☆ Apply chemical insecticide cartop hydrochloride 50 per cent SP @ 1000g/ ha or fipronil 80 per cent WG 62.5 g/ha or chlorpyriphos 20 per cent EC 1250 ml/ha or flubendamide 20 per cent WG 250 g/ha

Mealy Bug

The infestation starts in the plants one or two month after transplanting. Stunted, circular patches may be seen in the fields. If such plants are pulled out and teased the insects can be seen at the base of the leaves and leaf sheaths. Large number of these insects remains inside the leaf sheaths and suck up the plant sap. The affected tillers remain stunted with yellowish curled leaves. When the attack is severe, it inhibits panicle emergence. This type of disease is called as Soorai disease in Tamil Nadu. In severe cases, yield may be reduced even up to 50 per cent. Recently, the incidence of mealy bug has been reported during samba season in Thiruvarur and Nagapatinam districts. The newly hatched nymphs crowded within the waxy threads and body gets covered with waxy material on second day. Nymphs and adults being wingless look alike. Females are reddish, oval, soft-bodied living in colonies inside the leaf sheath and it reproduces parthenogenetically.

Management

- ☆ Remove the grasses form the buds and trim the bunds during the main field preparation before transplanting. Remove and destroy the affected plants.
- ☆ Spray any one of the following insecticides in the initial stage of infestation, fenitrothion 50 per cent EC1000ml, phosalone 35 per cent EC 1500 ml/ha.

Black Bug

The pest appears at the later stage of the crop and present at the base of the stem just above the water level. Plants become stunted with reduced number of tillers, leaves turn reddish brown and dry. The bugs remain and feed the plant sap on the base of the plants causing stunting of plants. Leaves turn reddish brown and grains do not develop. Bugs feed on the panicles in milky stage result in brown spots or empty grains in the panicles. Heavy bug infestation may cause death to the plants and whole field appears burned called bug burn similar to hopper burn. Young nymph is brown with yellowish green abdomen and a few black spots. Adults are flat, brownish black bugs with a prominent scutellum and pronotum having a spine on either side. It is active on the cloudy days and during night.

ETL

10 per cent damage at tillering stage or 5 bugs/hill.

Management

- ☆ Keep the field free from weeds and grasses
- ☆ Drain excess water from the filed
- ☆ Set up light trap to attract large number of bugs and kill
- ☆ Ducks can be allowed in the field to pick up the bugs
- ☆ Spray 5 per cent neem seed kernel extract 25 kg/ha for effective control of black bug

Ear Head Bug/Gundhi Bug

This insect pest occurs during panicle initiation stage. The bugs suck the sap from the developing grains at milky stage and appearance of numerous brownish spots at the feeding sites and grains become shrivelled. In the case of heavy infestation, the whole ear head may become devoid of mature grains. Its presence in the field is made out by its strong smell. Both adults and nymphs do the damage. Serious infestation can reduce the yield by 50 per cent. The straw gives off-flavour that is unattractive to cattle. Nymphs are small, pale green in colour, whereas adults greenish yellow, long and slender with a characteristic buggy odour.

ETL

5 bugs/100 panicles during flowering stage, 16 bugs/100 panicles at milky stage

Management

☆ Remove weeds and grasses from the field

☆ Avoid following ratoon crop as the ratoon crop crops serves as breeding host for the insect

☆ Dust quinalphos 1.5 per cent D at 25 kg/ha twice, the first during flowering and second a week later

☆ Spray 5 per cent neem seed kernel extract 25 kg/ha

☆ Spray insecticide malathion 50 per cent EC 500 ml/ha

Gall Midge

This pest occurs in irrigated and wet season crops. The symptoms appear as the central shoot instead of producing leaf produces a long tubular structure. When the gall elongates as an external symptom of damage, the insect will be in pupal stage and ready for emergence. The maggot bores into the growing point of the tiller and causes abnormal growth of the leaf sheath, which becomes whitish tubular and ends bluntly. It may be pale green, pink or purplish. Further growth of tiller is arrested. This is called onion shoot, silver shoot or anaikomban. The adults fly is yellowish brown and mosquito like.

ETL

10 per cent silver shoots

Management

☆ Destroy Alternate weed hosts such as *Cynodon dactylon, Elusine indica, Brachyieria mutica, Panicum* spp.

☆ Use resistant varieties like MDU 3, Vikram, Sureka

☆ Set up light trap as a monitoring device

☆ Conserve and enhance the activity of *Platygaster oryzae* in the main field by avoiding broad spectrum insecticides

☆ Spray chlorpyriphos 20 per cent EC 1200 ml (or) fibronil 5 per cent SC 1000 ml (or) thiomethaxam 25 per cent WG 100g/ha

Rice Case Worm

It is a occasional pest and its damage can be distinguished from damage by other pests in two ways, firstly the ladder like appearance of the removed leaf tissue resulting from the back and forth motion of the head during feeding and secondly the damage pattern is not uniform through out the field because the floating cases are often carried in the run off water to low lying fields where the damage is more concentrated. The plants stunted, caterpillars hanging on the leaf edges in a tubular case. The caterpillar cuts a piece of leaf, rolls it longitudinally into a tubular structure and remains inside. It feeds by scraping the green tissue of the leaf. The cases often float in the water.

Management

☆ Drain the water from the field to kill pupal cases

☆ Dislodge the larval case by running a rope over the young crop, mix kerosene 200 ml in water to kill the fallen larva

☆ Spray of phenthoate 50 per cent EC 1000 ml (or) carbaryl 10 per cent DP 25kg/ha

General Integrated Pest Management Practices in Rice

☆ Remove/destroy stubbles after harvest and keep the field free from weeds

☆ Trim and plaster the bunds of rice field to expose the eggs and various stages of insects to eliminate breeding in grasses

☆ Form the bunds narrow and short to reduce the damage by rodents

☆ Use resistant varieties wherever available

☆ Provide effective drainage wherever the problem of BPH

☆ Clip the tip of the seedling before transplanting to prevent the carry over of egg masses of rice yellow stem borer from nursery to main field.

☆ Organize synchronized planting wherever possible

☆ Grow cowpea in the field buds of main field to encourage natural enemies population

☆ Leave 30 cm rogue space at every 8 feet to reduce damage by insects and also for proper plant protection operation measures

☆ Avoid use of excessive nitrogenous fertilizers

☆ Install bird perches to encourage bird predators

☆ Set up light traps at early night hours to monitor and control insect pests

☆ Set up bow traps to kill rodents

☆ Alternate wetting and drying in the main field, Drain excess water from the field for the management of BPH, case worm insects

☆ Set up owl perches to reduce rat damage

☆ Avoid broad spectrum insecticides to protect the beneficial insects like spiders, dragonfly, damselfly, lady bird beetles, mirid bugs, ground beetle, rove beetles and minute parasitoids

☆ Use plant based products like neem seed kernel extract 5 per cent or neem oil 3 per cent

☆ Use chemical insecticides only when insect pests damage crosses the economic threshold level (ETL)

☆ Avoid use of resurgence causing insecticides and synthetic pyrethroids

☆ Use the recommended dose insecticides and avoid using over dose or sub-lethal dose of chemicals for the management of insect pests

Chapter 11

Integrated Disease Management in Rice

S. Nakkeeran, M. Karthikeyan and P. Renukadevi

Department of Plant Pathology,
Tamil Nadu Agricultural University,
Coimbatore – 3, Tamil Nadu
e-mail: nakkeeranayya@gmail.com

Rice production in Tamil Nadu is limited to the extend of 20-30 per cent by various diseases. The rice crop is infected by differents diseases during *kuruvai* (June–September), *samba/thaladi* (August–January) and summer (February–May) seasons in the state. However, the disease incidence is more *in samba/thaladi* season. Rice crop is affected by more than 50 diseases caused by fungi, bacteria and viruses. Among them, blast, sheath rot, sheath blight, brown spot, grain discolouration, false smut, bacterial leaf blight and bacterial leaf streak are the important diseases affecting the rice crop. Introduction of high yielding rice varieties/hybrids, excessive application of chemical fertilizers and climatic change are leading to severe occurrence of various diseases. In this juncture, adoption of IDM strategies will be an effective tool for the management of rice diseases.

FUNGAL DISEASES

Blast

The blast disease is caused by a fungus *Pyricularia grisea*. It causes yield reduction to the extent of 50–60 per cent. In case of neck blast, the yield loss will be up to 80 per cent. Normally the disease incidence is more during winter season and rainy period. The disease incidence is reported even in summer in susceptible varieties and hybrids. The disease causes damage to the crop both in nursery and mainfield. The disease is classified into two categories namely leaf blast and neck blast. Initially the disease appears as small brown specks on leaves and sheaths.

The spots enlarge into spindle shaped (eye shaped) spots with brown margin and greyish centre. As the disease advances, the spots are coalesce together and the leaves become dried. In the nursery, affected seedlings are completely blighted and become dead.

The pathogen infects on stem nodes and neck portion of the inflorescence as brown or black spots. The affected tissues become rotten and the panicle breaks at neck portion. Hence, it is called "Neck blast". Grains are poorly developed and become chaffy. The disease spreads rapidly when night temperature falls below 20°C and the relative humidity is about 90 per cent. Cloudy weather encourage spore germination and infection. Excessive application of nitrogenous fertilizer favours the disease incidence. Weed hosts like *Panicum repens, Brachiaria mutica, Echinochloa crusgalli, Digitaria marginata* on field bunds and irrigation channels serve as source of infection. The disease is also spread through seeds and air.

Wet seed treatment with Carbendazim or Tricyclozole or Thiram @ 2g/kg of seed in one litre of water for more than 12 hours was effective in reducing the inoculum. Spraying of Carbendazim 50 g/20 cent nursery is recommended for disease control. Destruction of weed hosts in the field bunds and irrigation channels are helpful to prevent the disease spread. In the mainfield, spraying of any one of the fungicides Carbendazim 250 g or IBP 500 ml or Tricyclozole 500g in 500 litres of water for one hectare is suggested for effective disease control.

Talc formulation of bio control agent, *Pseudomonas fluorescens* (Pf1) can be used for seed treatment @ 10 g/kg seeds instead of fungicides. Application of *Pseudomonas* at 1.5 kg/20 cent nursery along with 30 kg Farm Yard Manure (FYM), 48 hrs before pulling out the seedlings is advocated. In the transplanted field, the *Pseudomonas* can be applied @ 2.5 kg/ha along with 50 kg FYM at 30 days after transplanting. Application of plant growth promoting rhizobacteria *P. fluorescens* in the nursery and main field induces host resistance by producing antibiotics. As soon as the disease symptoms noticed, spraying of 2.5 kg *Pseudomonas* in 500 litres of water for one hectare will be effective in containing the disease.

Sheath Rot

Sheath rot disease is caused by a fungus *Sarocladium oryzae*. The disease infection is confined to the uppermost leaf sheath enclosing the panicle (flag leaf sheath). It causes considerable yield reduction to the extent of 60-80 per cent. Since the disease attacks at boot leaf, the yield loss is very high under favourable climatic condition. The disease incidence is found in all the seasons. Initial symptoms of the disease are appearance of oblong or irregular lesions measuring 1-2cm long with brown margin and grey centre. As the disease advances, the lesions enlarge, coalesce together and cover most of the leaf sheath area. This leads to choking and failure of panicle exertion. Depending upon the stage of infection, complete or partial exertion of the panicle is observed. The pathogen proliferates as white mycelium inside the sheath and the panicle become rotten. The pathogen also causes damage to the grains and make them brown and chaffy and serve as source of contaminant and primary source of inoculum.

Low temperature of 25-30°C and moderate relative humidity (75-80 per cent) favour the disease development. Application of high dose of nitrogenous fertilizer, adoption of closer spacing and infestation by sucking pests, stem borer aggravate the disease incidence. Insect damage pave way for the entry and spread of the pathogen. The pathogen is mainly spread through seeds and wind.

The following integrated disease management practices are recommended for containing the disease. Selection of healthy seeds and wet seed treatment with 0.2 per cent Carbendazim (soaking 1 kg seed in 1 litre of water mixed with 2g Carbendazim) for more than 12 hrs is useful to reduce seed borne inoculum. Soil application of gypsum at 500 kg/ha in two equal splits at basal and active tillering stage are useful. Spraying of fungicides Carbendzaim at 250 g/ha or Mancozeb 1.0 kg/ha at boot leaf stage and 15 days later reduce the disease severity. Foliar spray with neem oil 3 per cent or neem seed kernel extract 5 per cent is also effective. The bio control agent *P. fluorescens* can be applied as seed treatment at 10 g per kg of seed and 0.5 per cent foliar spray at the time of disease appearance for better disease management.

Sheath Blight

Disease is caused by *Rhizoctonia solani*. It causes spots or lesions mostly on the leaf sheath, extending to the leaf blades under favourable conditions. The spots are greenish grey at first, ellipsoid or ovoid and about 1cm long. They enlarge and may reach 2 or 3 cm. in length and become grayish white with brown margins and somewhat irregular in outline. In the advanced stages brown sclerotia are formed, which are easily detached from theses spots. Under humid conditions, the fungal mycelium may spread to other leaf sheaths and blades. Eventually the whole sheath rots and the affected leaf sheaths and blades. Eventually the whole sheath rots and the affected leaf can easily be pulled off, from the plant. In severe cases all the leaves of a plant are blighted resulting in death of the plant. Plants are usually attacked at the tillering stage, when the leaf sheaths become discoloured at or above water level.

The hyphae of the pathogen are hyaline when young, yellowish brown when old. Dark brown to brown, more or less globose sclerotia are produced on the media. The sclerotia are flattened on the lower side.

The fungus grows over a wide range of temperature, the optimum ranging from 28-30°C. The size and the number of sclerotia on culture media depend on the nitrogen source and concentration present therein.

The sclerotia of the fungus survive in the soil for several months depending on temperature and moisture conditions. The sclerotia float to the surface of the water during soil puddling, leveling, weeding and other operations and infect the plants with which they come into contact. The mycelium grows inside the tissues in all directions, initiating secondary spots, in turn producing sclerotia on the spots. The mycelium enters into the host plant through the stomata or it penetrates directly through the cuticles. The mycelium is most active and infectious when the lesions are young.

The disease is destructive during high humid and warm temperatures. Close planting and heavy fertilization tend to increase the incidence of the disease. High rates of nitrogenous fertilizer also make the tissue more susceptible to the disease while high potassium induces resistance to the disease. The intensity of the primary infection is closely related to the number of sclerotia that come into contact with the plant and subsequent disease development is greatly influenced by environmental conditions and susceptibility of the plants.

Management

Basal application of neem cake 150 kg/ha, followed by spraying of Propiconazole @500 ml or Carbendazim 250 g in 500 lit of water/ha is effective. Besides, spraying of *Pseudomonas* 0.5 per cent at the time of disease appearance, coupled with foliar application of neem oil 3 per cent or NSKE 5 per cent is also effective.

Brown Spot

The disease is caused by *Helminthosporium oryzae*. The disease causes blighting of seedlings. In grown up plants leaf spotting is the most common and readily obsvered symptom of this disease. The spots on the leaves and leaf sheaths are brown, round to oval, measuring about 0.5 to 2mm x 2 to 5mm. They are usually isolated but, in severe cases, may coalesce to form large patches of withered tissues. The grains also become infected and the black or dark brown spots on glumes are covered by olicaceous velvety growth. At times the neck region may be infected, causing symptoms similar to those of neck blast caused by *P. oryzae*. However, the neck region is brown or grayish brown as against blackening in blast, and the tissues are less weakened; hence the ear never breaks.Sometimes healthy grains harbour the fungus in the interior region. Only a low percentage of germination is obtained with the infected grains and often post-emergence seedling blight is observed as an unnoticed phase of this disease. Under highly favourable conditions the disease may be so severe as to cause a reduction in yield up to 90 per cent.

The fungus over winters mainly in infected plant parts. It is not soil-borne. Diseased seed may give rise to the seedling blight, the first phase of the disease. The young seedlings express symptoms soon after germination; pale yellowish-brown spots appear on the coleoptile, spreading to cover the other tissues of the seedling. This disease occurs naturally on as many as 20 different wild species of Oryzae. A few collateral hosts like *Digitaria sanguinalis, Leersia hexandra, Echinochloa colona. Pennisetum typhoides, Setaria italica* and *Cynodon dactylon*, on which the fungus is recorded through artificial inoculation, may serve as sources of primary inoculum. Secondary infection is by the first formed spores on the seedlings, which become wind-borne. The optimum temperature for the germination of the conidia is between 25 and 30°C and infection occurs when the humidity percentage is around 90 or above. Spread of the fungus inside the host is reported to be greater in darkness than in direct sunlight. Rice plants are most susceptible to the disease at the flowering stage. Heavy and late north-east monsoon and cloudy days appear to favour the disease. Nitrogenous manuring of the crop aggravates the disease incidence.

Seed treatment with *Pseudomonas* and spraying of Carbendazim 250 g or Mancozeb 1kg/ha when disease reaches grade 3 will contain the disease.

Grain Discolouration

This is an another major disease affecting the yield of rice hybrids and the quality of marketable produce. Germination of hybrid rice seeds is also affected. Grain discolouration was recorded in hybrid rice to the extent of 30-40 per cent. In Tamil Nadu, seed samples collected from different locations revealed that 14 genera of fungi were associated with grain discolouration. Among them, the most predominant were *Helminthosporium oryzae, Sarocladium oryzae, Alternaria tenuis, Trichoconiella padwickii, Fusarium moniliforme* and *Curvularia* sp. The per cent grain discolouration varied with variety and locations. Incidence of grain discolouration is more during wet season (Oct-Jan, Feb) in Tamil Nadu. Grain discolouration can be effectively contained by spraying of Carbendazim + Thiram (1: 1) 0.2 per cent at flowering and milky stages.

False Smut

False smut is caused by a fungus *Ustilaginoidea virens*. It is known as Lakshmi disease of rice. Presence of the disease was believed to be an indication of a bumper crop in the year. Climatic conditions favourable for the growth of the fungus are similar to the conditions of good crop growth. However, under congenial conditions, it is known to cause considerable yield loss. Yield loss is not only due to the occurrence of the smut balls but also due to increased sterility of kernels adjacent to the smut balls. Usually, only a few grains in a panicle is infected. At severe infection, most of the grains in the panicles are affected by the disease. The disease incidence is not only reducing the yield; but also affects the quality of grains or seeds. The prominent high yielding rice varieties like CO 43, CR 1009, ADT 38, ADT 39 and BPT 5204 are found to be infected by the disease. Many rice hybrids are also reported to be susceptible to the disease.

Due to fungal infection, individual grains of the panicles are transformed into greenish spore balls of velvetty appearance. Spore balls are small at first and are visible in between the glumes, growing gradually to reach one cm or more in diameter and enclosing the floral parts. They are slightly flattened, smooth, yellow and are covered by a membrane. The membrane bursts as the result of further growth and the colour of the ball becomes orange and later yellowish – green or greenish – black. At this stage, the surface of the ball cracks. When cut open, it is white in the centre, and consists of tightly woven mycelium together with the glumes and other tissues of the host.

The fungus produces chlamydospores which are formed on the spore balls. Chlamydospores germinate by germ tubes and form conidiophores bearing 1-3 conidia at the tapering apex. Conidia are ovoid and very minute. Some of the green spore balls develop one to four sclerotia in the centre. These sclerotia over winter in the field and produce stalked stromata in the successive cropping season. It forms a structure 'Perithecium' at the top which contains asci and ascospores.

There are two types of infection. One type takes place at a very early stage of flowering, when the ovary is destroyed, but the style, stigmas and anther lobes remain intact and are ultimately buried in the spore mass. Second type takes place when the grain is at matured stage. Spores accumulate on the glumes, absorb moisture, swell, and force the lemma and palea apart. The fungus finally contacts the endosperm and the growth is greatly accelerated. Ultimately the whole grain is replaced and enveloped by the fungus.

The fungus survives during off season by means of sclerotia as well as chlamydospores. It is believed that primary infection is initiated mainly by ascospores produced from the sclerotia. Chlamydospores play an important role in secondary infection, which is a major part of the disease cycle. Chlamydospores are air-borne, but do not be free very easily from the smut balls because of the presence of a sticky material. High moisture or rainfall accompanied by cloudy days during the period between flowering and maturity of grains favour the development of false smut. Plants grow under conditions of high fertility, favourable for the good growth of rice were more susceptible to the disease. Weed plants *Digitaria marginata*, *Panicum* sp. and wild species of *Oryza* are known to serve as alternate hosts.

False smut can be effectively managed by adopting integrated disease management practices. Healthy seeds or seeds free from sclerotia should alone be used. Early planted rice crop has less smut balls than the late planted crop. At the time of harvesting, diseased plants should be removed and destroyed so that sclerotia do not fall in the field. This will reduce primary inoculum for the next crop. Field bunds and irrigation channels should be kept clean to eliminate alternate hosts. Excess application of Nitrogenous fertilizer should be avoided. Regular monitoring on disease incidence during *rabi* season is very essential. Spraying of Copper hydroxide @ 1.25 kg/ha or Propiconazole @ 1.0 ml/l at boot leaf and milky stages will be more useful to prevent the fungal infection.

BACTERIAL DISEASES OF RICE

Bacterial Leaf Blight

Incidence of BLB is found to occur in all the seasons. It is severe during samba/thaladi (*rabi*) season. It affects the crop at all stages of its growth, but it is more prevalent at tillering stage. It is caused by a bacterium *Xanthomonas oryzae* pv. *oryzae*. Initially, the disease appears as water soaked lesions on the tip and edges of leaves. The lesions are developed in to elongated streaks with wavy margin. As the disease intensifies, the leaves are dried at the margins initially and then completely dried. The central portion of the affected leaf is seen as straw coloured. Affected plants exhibit withering of leaves stunted growth and appear as sunken patches in the field. It is called as kresek stage. When the disease occurs at maturity phase, yield reduction is very low.

Bacterial Leaf Streak

Bacterial leaf streak is commonly noticed in *rabi* season. The disease is incited by a bacterium *Xanthomonas oryzae* pv. *oryzicola*. Small, longitudinal, water soaked, brown colour, discontinuous streaks are found between the veins of infected leaves. When the infected leaves are seen against sunlight, yellow streaks found with alternate green patches. Streaks are coalesced together; the leaves become orange or reddish brown and subsequently dried. Affected plants show burnt appearance in the field. The disease spread through seed, rain water splashes and infected straw material.

The bacterial diseases can be effectively controlled by spraying 20 per cent fresh cow dung extract at initial stage. (40kg fresh cow dung is dissolved in 100 litres of water and kept overnight. The supernatant solution is made up to 200 litres of spray fluid with water which is required for one acre). Spraying of *P. fluorescens* (Pf 1) @ 400 g/ac controls the disease. Antibiotics streptomycin sulphate + oxy tetracycline at 300g/ha along with copper oxy chloride 1.25 kg or copper hydroxide 1.25 kg/ha is also effective.

Regular monitoring, proper diagnosis of disease incidence and adoption of integrated disease management practices at appropriate stages will be helpful to contain the diseases and to increase the production of rice.

Chapter 12

Nutritional Deficiency and its Management in Rice

D. Vijayalakshmi[1] and C. N. Chandrasekhar[2]

[1]Assistant Professor, [2]Professor and Head
Department of Crop Physiology,
T.N.A.U., Coimbatore, Tamil Nadu
e-mail: vijiphysiology@gmail.com

Introduction

The term nutritional disorder is refered as non-pathogenic and nutrient-related disorders which may be deficiencies or toxicities of an element or substance. The growth of the rice plant in any medium (soil, sand, water) depends on the availability of sunlight, water, and various chemical elements. Seventeen elements are recognized as essential nutrients in rice nutrition. They are Carbon, Hydrogen, Oxygen, Nitrogen, Phosphorus, Potassium, Calcium, Magnesium, Sulphur, Iron, Manganese, Copper, Boron, Zinc, Molybdenum, Chloride and Nickel. Among these, Carbon, Hydrogen, and Oxygen are absorbed directly from air and water while, the rest are present in the soil.

These nutrients are classified as - Nitrogen, Phosphorus and Potassium are known as primary plant nutrients; Calcium, Magnesium and Sulphur, as secondary nutrients; Iron Manganese, Copper, Zinc, Boron, Molybdenum and Chlorine as trace elements or micro-nutrients.

The primary and secondary nutrient elements are known as major elements. This classification is based on their relative abundance, and not on their relative importance. The micronutrients are required in small quantities, but they are important as the major elements in plant nutrition.

In considering the effects of individual elements, the relative amounts of other elements present is also important. For example, Nitrogen alone produces certain effects, but the effects may be quite different if there is a proper balance between Nitrogen, Phosphorus, Potassium, and other elements. The chemical form in which a nutrient element is present in the soil is also important, since the availability of a nutrient to the plant varies with the roots ability to extract the nutrient element from the chemical compound in which it occur.

Classification of Nutrients Based on Mobility of Nutrients in Plant System

Mobile Nutrients

- ☆ Nitrogen
- ☆ Phosphorus
- ☆ Potassium
- ☆ Magnesium

Deficiency symptoms appear in oldest (lower) leaves first, because their mobile nutrients contents move to the youngest leaves, which act as sinks.

Immobile Nutrients

- ☆ Calcium
- ☆ Iron
- ☆ Manganese
- ☆ Zinc
- ☆ Sulfur

Deficiency symptoms appear in youngest (upper) leaves first, because these nutrients become part of the plant compounds.

Role, Symptoms and Corrective Measures of Nutrients Deficiencies in Rice

Nitrogen (N)

Role

- ☆ Nitrogen encourages the vegetative development of plants by imparting a healthy green colour to the leaves.
- ☆ Promotes rapid growth (plant height and tiller number) and increased leaf size, spikelet number per panicle, percentage filled spikelets in each panicle, and grain protein content.

Symptoms

- ☆ Plants are stunted with limited tillers. Except for young leaves which are greener, leaves are narrow, short, erect, and yellowish green.
- ☆ Old leaves die when light straw coloured.

Plate 12.1: Symptoms of Nitrogen Deficiency.

Corrective Measures

- ☆ Soil application of 25 per cent excess of recommended N.
- ☆ Foliar application of Urea 1 per cent at weekly interval till the symptoms disappear.

Phosphorus (P)

Role

- ☆ Phosphorus is particularly important in early growth stages.
- ☆ It is mobile within the plant and promotes root development (particularly the development of fibrous roots),tillering and early flowering
- ☆ P deficiency is often associated with other nutrient disorders such as iron toxicity and low pH

Plate 12.2: Symptoms of Phosphorus Deficiency.

☆ It also increases resistance to disease and strengthens the stems of rice plants, thus reducing their tendency to lodge.

Symptoms

☆ Plants are stunted with limited tillers.

☆ Leaves are narrow, short, erect, and dirty dark green.

☆ Young leaves are healthy and old leaves die when brown coloured.

☆ A reddish or purplish color may develop on the leaves if the variety has a tendency to produce anthocyanin pigment.

Corrective Measures

☆ Application on P fertilizer 15-30 kg P/ha.

☆ Rock phosphate broadcast before flooding when soil pH is low.

☆ Application of phosphobacteria to the soil as seed coating or as seedling dip.

Potassium (K)

Role

☆ Potassium enhances the ability of the plants to resist diseases, insect attacks, cold and other adverse conditions.

☆ It plays an essential part in the formation of starch and in the production and translocation of sugars in rice.

☆ Delays leaf senescence

Plate 12.3: Symptoms of Potassium Deficiency.

☆ Improves root growth and plant vigor and helps prevent lodging.

☆ Helps in proper uptake of other nutrients.

Symptoms

☆ Plants are stunted, but tillering is only slightly reduced.

☆ Leaves are short, droopy, and dark green.

☆ The lower leaves at the interveins, starting from the tip, turn yellow and, eventually, dry to a light-brown colour.

☆ Rusty brown spots appear on the tips of older leaves causing it to turn brown and dry up.

☆ Weak stem leads to lodging.

Corrective Measures

☆ Soil application of 25 per cent excess of recommended K.

☆ Foliar application of 1 per cent KCl at 15days interval.

Sulfur (S)

Role of S in Rice

☆ Involved in chlorophyll production

☆ Required for protein synthesis, plant function and structure

Plate 12.4: Symptoms of Sulfur Deficiency.

☆ Effect on yield is more pronounced when S deficiency occurs during vegetative growth.

Symptoms

☆ Yellowing of whole plant

☆ Chlorosis is more pronounced in young leaves

☆ Stunted growth

☆ Delayed maturity

Corrective Measures

☆ Incorporate straw instead of completely removing or burning it. About 40-60 per cent of the S contained in straw is lost during burning

☆ Applying 15-20 kg S ha^{-1} gives a residual effect that can supply the S needed for two subsequent rice crops.

Calcium (Ca)

Role

☆ Formation of cell wall and maintain structure

☆ Cell division and enlargement, Lipid metabolism

☆ It also promotes the activity of soil bacteria concerned with the fixation of free nitrogen or the formation of nitrates from organic forms of nitrogen.

☆ It is necessary for the development of a good root system.

Plate 12.5: Symptoms of Calcium Deficiency.

Symptoms

☆ The general appearance of the plant is little affected except when the deficiency is acute. The growing tip of the upper leaves becomes white, rolled and curled.

☆ In an extreme case, the plant is stunted and the growing point dies.

Corrective Measures

☆ Apply farmyard manure or straw (incorporated or burned) to balance Calcium removal in soils containing small concentrations of Calcium Fertilizers.

☆ Apply $CaCl_2$ or Ca containing foliar sprays for rapid treatment of severe Ca deficiency.

☆ Apply gypsum in Ca deficient high pH soils, *e.g.,* on sodic and high K soils.

☆ Apply lime on acid soils to raise pH and Ca availability.

☆ Apply pyrites to mitigate the effects of $NaHCO_3^-$ rich water on Ca uptake.

Magnesium (Mg)

Role

☆ Synthesis of Chlorophyll

☆ It is usually needed by plant in relatively small quantities. Hence its deficiency in the soil is experienced later than that of Potassium.

Plate 12.6: Symptoms of Calcium Deficiency.

Symptoms

- ☆ Height and tiller number are little affected when the deficiency is moderate.
- ☆ Leaves are wavy and droopy due to expansion of the angle between the leaf blade and the leaf sheath.
- ☆ Interveinal chlorosis, occurring on lower leaves, is characterized by an orangish-yellow color.

Corrective Measures

- ☆ Rapid correction of Mg deficiency symptoms is achieved by applying a soluble Mg source such as kieserite or Mg chloride.
- ☆ Foliar application of liquid fertilizers containing Mg (e.g.,MgCl 22 per cent)

Iron (Fe)

Role

- ☆ Catalyst for Chlorophyll synthesis

Symptoms

- ☆ Interveinal yellowing
- ☆ Entire leaves become chlorotic and then whitish.
- ☆ If the iron supply is cut suddenly, newly emerging leaves become chlorotic.

Corrective Measures

- ☆ Apply solid $FeSO_4$ (30 kg Fe/ha) next to rice rows or broadcast.
- ☆ Foliar applications of $FeSO_4$ (2-3 per cent solution) 2-3 applications at 2 week intervals.

Plate 12.7: Symptoms of Iron Deficiency.

Manganese (Mn)

Role

- ☆ It is an activator of nitrite reductase and many respiratory enzymes.
- ☆ Play vital role in Photosynthesis
- ☆ Photolysis of water and Evolution of O_2
- ☆ Functions with enzyme systems involved in breakdown of carbohydrates and nitrogen metabolism.
- ☆ Soil is the source of manganese.

Symptoms

- ☆ Interveinal necrosis
- ☆ Plants are stunted but have a normal number of tillers.
- ☆ Interveinal chlorotic streaks spread downward from the tip to the base of the leaves, which later become dark brown and necrotic.
- ☆ The newly emerging leaves become short, narrow, and light green.

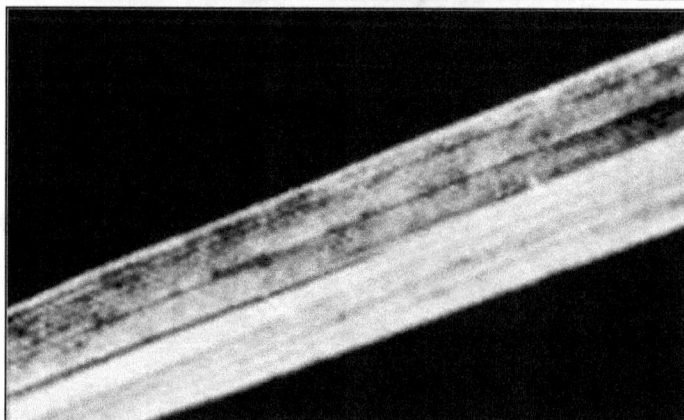

Plate 12.8: Symptoms of Manganese Deficiency.

Corrective Measures

☆ Apply MnSO₄ or finely ground MnO (5-20 kg Mn/ha) in bunds along rice rows.

☆ Apply foliar MnSO₄ for rapid treatment of Mn deficiency(1-5 kg Mn/ha in about 200 L water/ha).

Zinc (Zn)

Role

☆ Essential element in chlorophyll production in the rice plant

☆ Zn deficiency is the most wide-spread micronutrient-related problem in rice

☆ Essential for the transformation of carbohydrates.

☆ Regulates consumption of sugars.

☆ The function of zinc in plants is as a metal activator of enzymes.

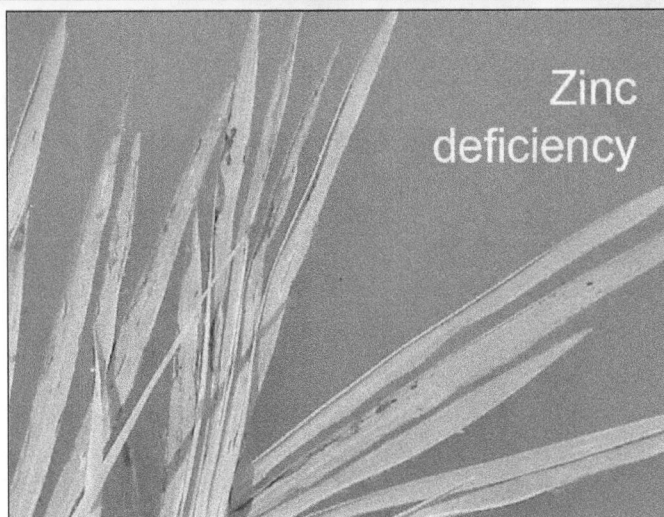

Plate 12.9: Symptoms of Zinc Deficiency.

Symptoms

- ☆ Dusty brown spots on old leaves
- ☆ Stunted plant growth
- ☆ Patches of poorly established hill in the field

Severe Zn Deficiency

- ☆ Decreases or stops tillering
- ☆ Increases time to crop maturity
- ☆ Increases spikelet sterility
- ☆ The midribs of the younger leaves, especially at the base, become chlorotic.

Corrective Measures

- ☆ Apply 25 kg of Zinc sulphate with 50 kg sand before transplanting.
- ☆ Apply 5-10 kg Zn ha^{-1} as Zn sulfate, apply 0.5 – 1.5 per cent $ZnSO_4$/ha as a foliar spray at tillering (25-30 DAT), 2-3 repeated applications at intervals of 10-14 days.

Boron (B)

Role

- ☆ Pollen germination and pollen tube growth
- ☆ Translocation of sugars and Hormone metabolism
- ☆ Cell division, differentiation and Development

Symptoms

- ☆ Young leaves are affected and death of shoot tip
- ☆ Abnormal growth, Irregular seed setting
- ☆ White and rolled leaf tips of young leaves.

Plate 12.10: Symptoms of Boron Deficiency.

☆ Reduction in plant height.

☆ Death of growing points, but new tillers continue to emerge during severe deficiency.

☆ Plants unable to produce panicles if affected by B deficiency.

☆ Plant height is reduced.

Corrective Measures

☆ Apply B in soluble forms (Borax) for rapid treatment of B deficiency (0.5-3 kg B/ha), broadcast and incorporated before planting, top dressed or as foliar spray during vegetative rice growth.

Copper (Cu)

Role

☆ It is an important constituent of plastocyanin (copper containing protein)

☆ Important constituent of enzymes and co-enzymes

☆ Important for reproductive growth.

☆ Aids in root metabolism and helps in the utilization of proteins.

Symptoms

☆ The leaves appear bluish green, and then become chlorotic near the tips.

☆ The chlorosis develops downward along both sides of the midrib; it is followed by dark-brown necrosis of the tips.

Plate 12.11: Symptoms of Copper Deficiency.

☆ The new, emerging leaves fail to unroll and appear needle-like for the entire leaf or, occasionally, for half the leaf, with the basal end developing normally.

☆ Reduced tillering and increased spikelet sterility

Corrective Measures

☆ Dip seedling roots in 1 per cent $CuSO_4$ suspensions for 1 hr before transplanting.

☆ CuO or $CuSO_4$ (5-10 kg Cu/ha at 5-year intervals) for long-term maintenance of available soil Cu (broadcast and incorporate in soil).

Silicon (Si)

Role

☆ Increases the availability of P, Ca and Mg

☆ Prevents Mn and Fe toxicity

☆ Keeps the leaves to erect and uncurved - Stiffness

☆ Prevents lodging –Mechanical strength

☆ Pest and disease resistance

☆ An adequate supply of silica is essential for paddy to give a good yield by increasing the strength and rigidity of the cells.

Symptoms

☆ Leaves become soft and droopy.

☆ Leaves and culms become soft and droopy thus increasing mutual shading.

☆ Reduces photosynthetic activity.

☆ Severe Si deficiency reduces the number of panicles m^2 and the number of filled spikelets per panicle.

☆ Si-deficient plants are particularly susceptible to lodging.

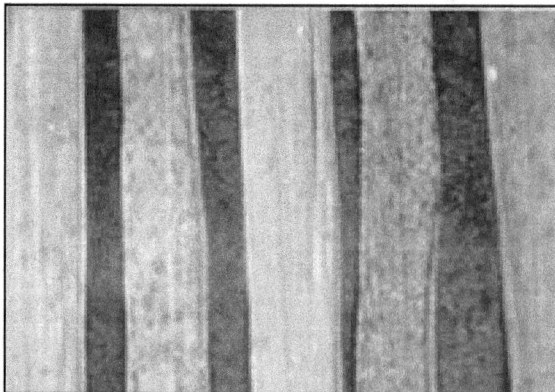

Plate 12.12: Symptoms of Silicon Deficiency.

Corrective Measures

☆ For more rapid correction of Si deficiency, granular silicate fertilizers should be applied:

Calcium silicate 120-200 kg/ha, Potassium silicate 40-60 kg/ha.

Chapter 13

Recent Physiological and Molecular Advances to Combat Major Abiotic Stresses in Rice

D. Vijayalakshmi[1] and P. Jeyakumar[2]

[1]Assistant Professor, [1]Professor and Head,
Department of Crop Physiology,
TNAU, Coimbatore, Tamil Nadu
e-mail: vijiphysiology@gmail.com; physiology@tnau.ac.in

Introduction

World population is escalating day by day and by 2050 it is expected to reach 9.1 billion, but agricultural production is not rising at a parallel pace. In order to feed the world population, global agricultural production should be increased by 60-110 per cent and 70 per cent more food for an additional 2.3 billion people by 2050 (Tilman *et al.*, 2011). Agriculture production is dwindled mainly due to biotic and abiotic stresses. Abiotic stress is one of the major factors which negatively affect the crop growth and productivity world-wide. Hence, this is one of the major area of concern to fulfill the required food demand. It can impose limitations on crop productivity and also limit land available for farming, often in regions that can ill afford such constraints, thus highlighting a greater need for understanding how plants respond to adverse conditions with the hope of improving tolerance of plants to environmental stress. Rice carries an odd portfolio of tolerances and susceptibilities to abiotic stresses as compared to other crops. Among these stresses, drought is the first environmental stress responsible for decrease in agricultural production worldwide. Rice thrives in waterlogged soil and can tolerate submergence at levels that would kill other crops, but every year several million hectares of rice area is subjected to submergence or flooding stress wherein rice plants are completely submerged for 1–2 weeks, resulting in partial or even complete crop failure. Rice is moderately tolerant of salinity and soil acidity, but is highly sensitive to drought and it suffers from low temperatures at seeding and high temperatures at flowering. Most rice is currently grown in regions where current temperatures are already close to optimum for rice production. Therefore, any further increases in mean temperatures or of short episodes of high temperatures during sensitive stages may be supraoptimal and reduce grain yield.

Drought affects plants in countless ways like, it affects plant growth, yield, membrane integrity, pigment content, osmotic adjustments, water relations and photosynthetic activity. Similarly, flash-floods are highly unpredictable and may occur at any growth stage of the rice crop and the yield loss may be anywhere between less than 10 and 100 per cent depending on factors such as water depth, duration of submergence, temperature, turbidity of water, rate of nitrogen fertilization, light intensity and age of the crop. Salinity is the most prevalent soil problem in rice-growing countries after drought and rice is considered as a salt sensitive crop in early seedling stages which limits its productivity. Excess salts adversely affect all major metabolic activities in rice including cell wall damage, accumulation of electron-dense proteinaceous particles, plasmolysis, cytoplasmic lysis and damage to ER, reduction in photosynthesis and overall decline in germination and seedling growth leading ultimately to reduced growth and diminished grain yield. Hence, the major challenge in today's research scenario is to incorporate knowledge of physiology, biochemistry and molecular biology to mitigate the abiotic stresses affecting rice productivity.

Physiological and Molecular Responses of Rice Plants to Drought Stress

Rice is sensitive to drought stress, particularly during flowering stage, resulting in severe yield losses. A major constraint leading to spikelet sterility under drought stress is the failure of the panicle to exert fully from the flag-leaf sheath. Other possible reasons for this adverse effect of drought include retention of the panicle inside the flag-leaf sheath. A given level of drought at the vegetative stage can cause a moderate reduction in yield, but the same stress can eliminate yield entirely if it coincides with pollen meiosis or fertilization (O'Toole 1982). Some lowland cultivars have impressive levels of tolerance of tissue water deficit (Lilley and Ludlow 1996) and perform well in screens for plant survival with vegetative stage stress.

Mechanisms Underlying Drought Stress Tolerance in Rice

The physiological basis of genetic variation in drought response is not clear, in part because so many different measures of tolerance have been reported. If tolerance is defined as the ability to maintain leaf area and growth under prolonged vegetative stage stress, the main basis of variation appears to be constitutive root system architecture and its associated tillering habit that allows maintenance of more favorable plant water status (Nguyen *et al.*, 1997). Differences have also been observed in the adaptive response of root distribution to soil drying (Azhiri-Sigari *et al.*, 2000; Liu *et al.*, 2004). The mechanisms underlying genetic variation in both constitutive and adaptive root distribution may be sensitivity to signals, particularly auxin, that influence root elongation and branching (Bao *et al.*, 2004; Ge *et al.*, 2004). When drought tolerance is defined as the ability to flower and produce grain under water deficit, additional mechanisms may become important. Delayed flowering under drought is associated with an apparent delay in floral development when stress occurs between panicle initiation and pollen meiosis (from 30 to 10 d before heading). With the onset of stress occurring from 10 to 5 d before heading, flowering is slowed mainly due to slower elongation of the panicle and supporting tissues. Genetic variation in flowering delay under drought has been reported, and only part of this variation depended on measured plant water status.

Drought also affects the process of starch deposition in pollen grains, which normally begins about 3 days before anthesis, contributing to reduced anther dehiscence. Genetic variation for the tolerance of anther dehiscence to low water status has been observed. Panicle desiccation can occur when drought coincides with heading, variety-specific mechanisms that can refill cavitated xylem elements in shoots may be important to limit panicle failure (Stiller *et al.*, 2003).

Molecular approaches to drought tolerance have been widely applied to rice, beginning with QTL analysis. The rice genetic map is well covered by microsatellite markers (McCouch *et al.*, 2003), and rice researchers worldwide have developed diverse mapping populations and related databases. Mapping studies have been successful in identifying genetic regions associated with highly heritable traits such as plant height and flowering date, and in some cases it has been possible to identify the specific gene underlying a QTL. QTLs have also been identified for some secondary traits that are expected to be associated with drought response, such as rooting depth, membrane stability and osmotic adjustment. The results of QTL studies will probably be most usefully applied to the identification of promising genetic regions for the identification of candidate genes. Nonetheless, modifications of QTL mapping strategies still hold promise to deliver a product that will be more directly useful for cultivar improvement – these include linkage disequilibrium studies and the use of the advanced-backcross QTL approach that combines selection and QTL identification in closely related backcross lines.

Recent advances in rice anatomy, physiology, and molecular biology have resulted in temporal and spatial protein and transcript expression profiling. With the availability of the complete rice genome sequence and rapid developments in the application of mass spectrometry in biological research, many tissue and organelle specific proteomic studies have been carried out to understand various physiological and molecular processes

Recent Strategies for Breeding Drought Stress Tolerance in Rice

Drought is the most serious constraint to rice production in unfavorable rice growing areas, and most of the popular farmers varieties are susceptible. Genetic studies on component traits have been carried out but these studies have generally not identified any QTLs that could be considered as promising targets in rice breeding programs. However, the use of direct measurement of yield under drought stress has shown more promising results. Bernier *et al.*, (2007) detected a QTL on chromosome 12 in a large population from the cross of Vandana/Way Rarem that accounted for about 50 per cent of the genetic variance, and was expressed consistently over 2 years. This QTL seems to be related to increased water uptake of plants under stress. A chromosome 3 QTL had a large effect on drought tolerance in the cross between the tolerant variety Apo and the widely grown susceptible variety Swarna (Venuprasad *et al.*, 2009). This is a very promising QTL for use in marker-assisted selection, because the variety Swarna is widely grown in drought-prone environments due to its high yield and other desirable traits. Promising varieties with drought tolerance like Sahbhagi Dhan (IR74371-70-1-1) have been developed through conventional breeding and are being disseminated to farmers in drought-prone areas. These varieties perform well even during favorable years, and they provide about 1 t/ha yield advantage under severe drought stress (Verulkar *et al.*, 2010). Most of the popular varieties grown by the farmers fail under these conditions.

Many studies report changes in the expression of individual genes when rice is challenged by drought stress, and they frequently respond to other abiotic and biotic stresses as well. These include diverse genes such as MAP kinase, DREB genes, calcium-dependent protein kinase, an endo-1,3-glucanase, a translation elongation factor (Li Zi and Chen Shou 1999), and glutathione reductase. Transformation studies have demonstrated that altering the expression of a number of different genes from different pathways can affect the response of rice to water deficit or dehydration. These include genes associated with diverse functions, such as water uptake (aquaporins), signaling (kinases), membrane integrity (LEA protein) and carbohydrate metabolism (TPS). The effect of transformation on grain production under stress has not been well documented. Because of the very large number of genes that change expression under drought stress, genomic approaches that can follow transcriptional changes in thousands of genes at a time hold great promise.

Physiological and Molecular Responses of Rice Plants to Flooding Stress

Rice cultivars differ in their tolerance to submergence and cultivars with adequate submergence tolerance do exist. FR13A cultivar from Eastern India, can survive for more than two weeks completely submerged. However, submergence tolerant cultivars are poor in yields, with low phenotypic acceptability by farmers due to inferior quality of grain. Understanding the physiological mechanism for cultivar differences in submergence tolerance may facilitate future breeding programmes.

Mechanisms for Improving Rice Crop Performance under Submergence Conditions

Gas Diffusion and Photosynthetic Rates

Limited gas diffusion is the most important factor during flooding. Since gas diffusion is 104-fold slower in solution than in air, the depletion of O_2 is a major feature of the flooded field which creates a condition of low O_2 (hypoxia) or no O_2 at all (anoxia) around the plant tissues such as seeds or root apices and stele, even if the latter contain aerenchyma. The maximum photosynthetic rates of rice cultivars decreased during seven days of complete submergence.

Sustained Sugar Supply and Energy Metabolism

High amounts of water soluble carbohydrates and starch prior to submergence and slower rates of depletion during submergence are adaptive traits. High initial stem carbohydrates and ability to sustain CO_2 assimilation during flooding are probably essential for providing energy for rapid growth to keep pace with the rising water. Initiation of new leaves and their subsequent growth requires availability of non-structural carbohydrates. Cultivars that maintained more than 6 per cent of their initial non-structural carbohydrate at the time of re-aeration were found to be capable of developing new leaves. High carbohydrate status after submergence is the consequence of its level before submergence. Seedlings of tolerant species have 30–50 per cent more non-structural carbohydrates compared to the susceptible cultivars. These non-structural carbohydrates are utilized during submergence to supply the required energy for growth and maintenance metabolism.

Shoot Elongation

Minimal underwater elongation during submergence is an important tolerance trait because elongating plants would tend to lodge as soon as the water level recedes. Rapid leaf elongation during submergence competes with maintenance respiration for carbon sources. Hence, minimum underwater elongation is associated with increased survival. The possible mechanism of submergence induced shoot elongation was mediated by an interaction between ethylene and GA. This negative relationship between submergence tolerance and elongation during submergence was confirmed using IRRI gene bank database of 903 cultivars (Setter and Laureles, 1996).

Hormonal Regulation

Plant hormones are known to play a role in survival under flooded conditions by regulating shoot elongation and leaf senescence. The fast growth of cultivars adapted to deep water conditions seems to be mediated by the action of three different plant hormones, ethylene, GA and ABA. Under anaerobic conditions, ethylene concentration increases in plant tissue because of both increased synthesis and entrapment. This causes a reduction in ABA concentration with a concomitant increase in GA level as well as responsiveness, resulting in enhanced shoot elongation. Conversely, tolerance of flash flooding is associated with a increased level or responsiveness to ethylene. Since enhanced stem elongation is not desired under flash flood conditions, blocking GA synthesis could enhance survival by reducing elongation and conserving energy for maintenance and regeneration growth during and after submergence, respectively (Setter and Laureles, 1996).

Alcoholic Fermentation

Submergence can shift aerobic respiration to the less efficient anaerobic fermentation pathway. This pathway depends on continued supply of substrate (glucose) and the two key enzymes, alcohol dehydrogenase (ADH) and pyruvate decarboxylase (PDC). Alcoholic fermentation is the key catalytic pathway for recycling NAD to maintain glycolysis and substrate level phosphorylation in the absence of oxygen. Increased alcoholic fermentation is thus one way to alleviate the adverse effect of anoxia on reduced production of energy for growth and maintenance purposes.

Recent Strategies for Breeding Flooding Stress Tolerance in Rice

Research shows that the mega varieties could be efficiently converted to a submergence tolerant variety in three backcross generations, involving a time of two to three years. Polymorphic markers for foreground and recombinant selection were identified to develop a wider range of submergence tolerant varieties to meet the needs of farmers in the flood-prone regions. This approach demonstrates the effective use of marker assisted selection for a major QTL in a molecular breeding program.

The basis of a marker-assisted backcrossing (MAB) strategy is to transfer a specific allele at the target locus from a donor line to a recipient line while selecting against donor introgressions across the rest of the genome. The use of molecular markers, which permit the genetic dissection of the progeny at each generation, increases the speed of the selection process, thus increasing genetic gain per unit time. The main advantages of MAB are: (1) Efficient foreground selection for the target locus, (2) Efficient background selection for the recurrent parent genome, (3) Minimization of linkage drag surrounding the locus being introgressed, and (4) Rapid breeding of new genotypes with favorable traits.

The effectiveness of MAB depends on the availability of closely linked markers and/or flanking markers for the target locus, the size of the population, the number of backcrosses and the position and number of markers for background selection. MAB has been successful towards transferring the *Sub1* loci conferring tolerance against submergence into the susceptible genotypes. Submergence tolerant versions of mega varieties like Swarna, Samba Mahsuri, Savithri, CR 1009 and IR 64 have been developed through MAB breeding.

Physiological and Molecular Responses of Rice Plants to Salt Stress

Rice is a salt-sensitive crop. Excess salts adversely affect all major metabolic activities in rice including cell wall damage, accumulation of electron-dense proteinaceous particles, plasmolysis, cytoplasmic lysis and damage to ER, accumulation of citrate, malate and inositol in leaf blades within 1 d of salt treatment, increase in proline levels by 4- to 20-fold, decrease in Fv/Fm ratio, reduction in photosynthesis and overall decline in germination and seedling growth (Yeo *et al.*, 1985; Lutts *et al.*, 1995; Garcia *et al.*, 1997; Khan *et al.*, 1997; Pareek *et al.*, 1997; Sivakumar *et al.*, 1998), leading ultimately to reduced growth and diminished grain yield. Rice is considered as more sensitive to salts during early seedling than at reproductive stages

Mechanisms Underlying Salt Stress Tolerance in Rice

For rice, soil salinity beyond ECe ~ 4 dS/m is considered moderate salinity while more than 8 dS/m becomes high. Similarly pH 8.8 - 9.2 is considered as non-stress while 9.3 to 9.7 as moderate stress and equal or greater than 9.8 as higher stress. (Here $pH_{1:2}$ is the pH of the solution with one part of soil and two part of distilled water). Though numerous physiological and biochemical changes take place under altered stress environment but only few of them change very significantly and also contribute a lot to the salt tolerance mechanism. These changes control the solute and water balance and their distribution on whole plant and tissue basis. Based on the studies it was observed that most of the rice varieties encounters following physiological and biochemical manifestation under higher salt stress conditions.

- ☆ High Na^+ transport to shoot
- ☆ Preferential accumulation of Na in older leaves
- ☆ High Cl^- uptake
- ☆ Lower K^+ uptake
- ☆ Lower fresh and dry weight of shoot and roots
- ☆ Low P and Zn uptake
- ☆ Change in esterase isozyme pattern
- ☆ Increase of non-toxic organic compatible solutes
- ☆ Increase in Polyamine levels
- ☆ Increase level of Reactive Oxygen Species (ROS)
- ☆ Quick response for the partial closure of stomata on salt-stress signaling

Seedling Vigour

Rice cultivars differ substantially in their growth rate with the most vigorous lines being the traditional varieties. Differences in vigor among rice cultivars accounted for much of the variation in survival under high salinity. Early seedling vigour is desirable due to high sensitivity during this stage coupled with high salinity levels encountered at the beginning of the season.

Initial Entry of Salts from Roots

Substantial genetic variability in the rate of sodium uptake by rice roots is present signifying a sizeable potential for genetic improvement. Plant roots experience the salt stress when Na^+ and Cl^- along with other cations are present in the soils in varying concentration (1 to 150 mM for glycophytes; more for halophytes). Ion uptake depends upon the plant growth stage, genotype, temperature, relative humidity and also light intensity. Excessive amount of salt in the rhizosphere retards the plant growth, limits yield and even cause the plant death.

Ion Homeostasis Pathway

Ion homeostasis in cell is taken care of by the ion pumps like antiporters, symporters and carrier proteins on membranes (plasma membrane or tonoplast membrane). Salt Overly Sensitive (SOS) regulatory pathway is one good example of ion homeostasis. Overexpressing of vacuolar transporter (NHX1) has increased the salinity tolerance of rice (Fukuda *et al.*, 2004).

Synthesis of Osmoprotectants

Most of the rice plants accumulate certain organic solutes such as sugar alcohol, proline, quarternary ammonium compounds in response of osmotic stress termed as compatible solutes because even in high concentration they do not interfere with enzymatic activities. Trehlose, a non-reducing sugar, possess a unique feature of reversible water storage capacity to protect biological molecules from desiccation damages. The increase in trehlose levels in transgenic rice lines of Pusa basmati 1 using either tissue specific or stress dependent promoter, resulted into the higher capacity for photosynthesis and concomitant decrease in the extent of photo-oxidative damage during salt drought and low temperature stresses.

Stress Activated Protein Pathway

Plants produce many kinds of stress responsive proteins induced by various kind of stresses like heat, cold, salt or drought etc. Major proteins are LEA and dehydrins etc. These are reported to play an important protective role during desiccation/salt stress in rice plants (Moons *et al.*, 1995).

Upregulation of the Anti-oxidant System

Salt stress in plants induce higher concentration of Reactive Oxygen Species (ROS)/intermediate such as superoxide, H_2O_2 and hydroxyl radicals due to the impaired election transport processes in chloroplast, mitochondria and photorespiration pathway. Under normal growth conditions, the production of ROS in cell in as low as $240MS^{-1}$ superoxide and the steady state level of H_2O_2 in chloroplast is 0.5M. However, under salinity, the level of ROS production reaches to as high as $720MS^{-1}$ (3 fold) and H_2O_2 level as high as 15M (30 fold). There is variability among rice genotypes for the enzymatic and non-enzymatic scavenging system hence it is possible to tag the genes coding for both enzymatic and non-enzymatic ROS scavenging agents and use them in engineering the desired plants.

Compartmentation of Ions in Older Leaves, Leaf Sheath and Culm

The ability of rice cultivars to compartmentalize ions in older leaves and structural tissues could crucially affect plant survival. Maintaining younger leaves at low salt concentrations probably contributes to the ability of certain varieties to survive saline conditions if they maintain their rates of leaf initiation at least equal to rates of leaf death. Rice cultivars differ in their ability to maintain the younger leaves at low sodium concentrations. Selection of plants on the basis of shoot appearance and/or whole shoot sodium content may not always reflect resistance in genotypes where this mechanism plays a major role in their salt tolerance.

Recent Strategies for Breeding Salt Tolerance in Rice

To maximize the productivity of rice under saline soils, there is an urgent need to look for: (1) sources of genetic variation that can be used for developing new cultivars with greater yield potential and stability over seasons and ecogeographic locations, (2) identification of molecular markers associated with salt stress tolerance genes or QTL conferring tolerance to salt stress conditions for their use in marker-assisted breeding programs, and (3) discovery of genes imparting salt tolerance and their introduction in salt-sensitive rice cultivars (Blumwald and Grover 2006).

Genetic Variation for Salt Tolerance

Source(s) of salt tolerance is unavailable within the cultivated germplasm of *O. sativa*. Nonetheless, there are clear indications that some traditional cultivars and landraces of rice (*e.g.*, Pokkali, Nona Bokra, Bura Rata, etc.) are more salt tolerant than many elite cultivars. In fact, Pokkali rice has been frequently used as a donor of salt tolerance genes in breeding programs. Extensive physiological and molecular studies have been carried out employing Pokkali rice for understanding the mechanisms of salt tolerance (Kawasaki *et al.*, 2001). The increased salt tolerance of Pokkali is usually attributed to both its capacity to maintain a low Na^+/K^+ ratio in shoot tissues and its faster growth rate under saline conditions. Na^+ is buffered in shoots of Pokkali by intracellular compartmentalization which is a basis of its robust salt stress response.

Molecular Markers and QTLs Linked to Rice Salt Tolerance

QTLs have been identified in rice for salinity tolerance/contributing traits and majority of them have been reported on chromosome 1. Major one of them are *saltol 1* (Gregorio *et al.*, 2002; Bonilla *et al.*, 1998), QTL for Na^+ uptake, K^+ concentration and Na/K ratio (Koyama *et al.*, 2001); *SKC1* or *OsHKT8, RNTQ1, SDS1*, Na^+ and Cl^- transport in stem and *qST1*. There are more reports of other big-effect QTLs on other chromosomes for the contributing traits and have been reported on Chromosomes 3, 4, 10 and 12; chromosomes 4, 6 and 9 (Koyama *et al.*, 2001); Chromosomes 4, 6, 7 and 9 (Lin *et al.*, 2004); Chromosomes 2, 3, 8, and 9. IRRI has already using marker assisted selection for *saltol* QTL, responsible for seedling stage salinity tolerance in rice, to facilitate the breeding for salt tolerant cultivars.

Salt Tolerance in Rice through Transgenic Approach

Transgenic rice plants tolerant to salt stress have been produced taking clues from the response of rice plants to salt stress, by restricting the uptake of salt and adjust their osmotic pressure by the synthesis of compatible solutes (proline, glycine betaine, sugars, etc.) and sequestering salt into the cell vacuoles for the maintenance of low cytosolic Na^+ levels. Enzymes that catalyze rate-limiting steps in the biosynthesis of compatible osmolytes and proteins that protect membrane integrity, control water or ion homeostasis, and bring about scavenging of reactive oxygen species (ROS) are the examples of stress tolerance effectors. Regulatory determinants include transcription factors that interact with promoters of osmotic stress regulated genes (like rd22BP1, AtMYB2, DREBIA, and DREB2A) and intermediates that post-transcriptionally activate the effectors [SOS3 (Ca^{2+} binding protein), SOS2 (a Ca^{2+} dependant protein kinase), and SOS1 (putative plasma membrane Na^+/H^+ antiporter).

Physiological and Molecular Responses of Rice Plants to High Temperature Stress

The optimum temperature for the normal development of rice ranges from 27 to 32 °C (Yin *et al.*, 1996). High temperature affects almost all the growth stages of rice, *i.e.* from emergence to ripening and harvesting. The developmental stage at which the plant is exposed to heat stress determines the severity of the possible damage to the crop. However, flowering (anthesis and fertilization) and to a lesser extent the preceding stage booting (microsporogenesis) are considered to be the stages of development most susceptible to temperature in rice. Temperatures higher than the optimum induced floret sterility and thus decreased rice yield (Nakagawa *et al.*, 2003). Spikelet sterility was greatly increased at temperatures higher than 35 °C.

Mechanisms Underlying High Temperature Stress Tolerance in Rice

Heat tolerance is generally defined as the ability of the plant to grow and produce economic yield under high temperature. As plants cannot move, the only option they have to defend themselves from various stresses is to make metabolic and structural adjustments. Rice tolerance classically comprises elements of escape or avoidance, *i.e.* the timing of panicle emergence and spikelet/floret opening relative to the occurrence of the stress, and the absolute tolerance of key processes, such as anther dehiscence, to the stress.

The ability of a variety to decrease its spikelet temperature with reduced RH can be considered as an avoidance mechanism, while the variability among the varieties in spikelet sterility at a given spikelet temperature could be considered as true tolerance. True heat tolerance at sensitive stages might be conferred by protecting structural proteins, enzymes and membranes from heat damage. The role of heat shock proteins (HSPs) and other stabilizing factors is crucial in these processes (Maestri *et al.*, 2002).

Acquired or induced thermo-tolerance is the ability induced by a sub-lethal heat stress to overcome subsequent exposure to lethal high temperatures. This type of tolerance is a cell autonomous phenomenon that results from an exposure to a short period of sub-lethal temperature or other moderate stress treatment (Larkindale and Vierling 2008). It depends mainly on the induction of specific pathways during the acclimation period and subsequent acquisition of thermo-tolerance. Quantitative trait loci (QTL) studies have shown that heat tolerance of rice during flowering is under polygenic control.

Plant Architecture and Time of Flowering and Anthesis

Plant architecture can play an important role in high temperature stress tolerance. If the plant morphology is such that the panicle is surrounded by many leaves, the plant will be able to withstand high-temperature stress due to increased transpirational cooling and by preventing evaporation from the anther due to its shading by the leaves. The reduced evaporation from the anther will ensure swelling of the pollen grains, an important trait for anther dehiscence.

The anthesis time during the day is important because spikelet sterility is induced by high temperature during or soon after anthesis (1–3 h after anthesis in rice), but not after fertilization is completed.

Length of Anther and Size of Basal Sore

It has been suggested that cultivars with large anthers are tolerant of high temperature at the flowering stage (Matsui and Omasa, 2002). As the direct cause of temperature stress is the reduction in the number of pollen grains that germinate on the stigmata. Pollen grains in the anthers with large basal pores would readily drop out of the basal pores on to the stigmata. In contrast, most of the pollen grains in anthers with small basal pores would remain in the anthers at the time of floret opening (Matsui and Kagata, 2003). Thus, this trait, along with the length of the anther, increases the chances of pollination, mainly by facilitating pollen release from the anther. As these characters can be easily identified compared with some other morphological traits, they can be used easily in breeding and as screening tool for selection of rice germplasms for high-temperature resistance.

Mechanism of Heat-Induced Floret sterility

A key mechanism of high-temperature induced floret sterility in rice is the decreased ability of the pollen grains to swell, resulting in poor thecae dehiscence. This swelling of pollen grains in the locules is the driving force for anther dehiscence (Matsui *et al.*, 1999). It is found that although high-temperature-treated pollen showed a normal round shape, some of the tapetum functions such as pollen adhesion to the stigma and its subsequent germination were negatively affected. Some other possible reasons for decreasing spikelet fertility at high temperature are altered hormonal balance in the floret, disturbance in the availability and transport of photosynthates to the kernel, lack of ability of the floral buds to mobilize carbohydrates under heat stress and changes in the activities of starch and sugar biosynthesis enzymes. Greater increments in temperature resulted in higher proportions of sterility. Exogenous application of growth regulators has been shown recently to have some positive effects on the spikelet fertility and pollination. In fact, their exogenous application increased the level of endogenous antioxidants and thus prevented the oxidative damage to the membranes in rice. Humidity also plays an important role in rice yield, as higher relative humidity (RH) at the flowering stage under increased temperature affects spikelet fertility negatively. Decrease in the fertility of spikelets at high air temperatures with increased humidity was because humidity modified the impact of high temperature on spikelet fertility. An RH of 85–90 per cent at the heading stage induced almost complete grain sterility in rice at a day/night temperature of 35/30°C.

Pollination in Relation to Spikelet Fertility at High Temperature

Generally, male reproductive development in rice is known to be more sensitive to heat stress. High-temperature stress during rice flowering led to decreased pollen production and pollen shed. The probable reasons were the inhibition of swelling of pollen grains, indehiscence of anthers and poor release of pollen grains, and thus fewer numbers of pollen grains were available to be intercepted by the stigma.

Physiologically, the decreased production of pollens at elevated temperatures may be attributable to impaired cell division of the microspore mother cells. Similarly, high temperatures at anthesis or soon after can cause poor pollen germination and retarded pollen tube growth, along with poor anther dehiscence. Exposure of pollen grains to high temperature resulted in a loss of pollen viability within 10 min (Song *et al.*, 2001) and it was essential that more than 10 pollen grains germinated on the stigmata to ensure successful fertilization of a rice floret

Role of HSPs in Inducing Thermo-tolerance

High temperatures generally induce the expression of HSPs and suppress, at least in part, the synthesis of normal cellular proteins. The rapid accumulation of HSPs in the sensitive organs can play an important role in the protection of the metabolic apparatus of the cell, thereby acting as a key factor for plants' adaptation to and survival under, heat stress (Wahid *et al.*, 2007). These HSPs can help in coping with heat stress by improving photosynthesis, partitioning of assimilate, nutrient and water use efficiency and membrane thermal stability. A positive relationship has been documented in many plant species between HSPs and heat tolerance of the whole plant.

Recent Strategies for Breeding High Temperature Stress Tolerance in Rice

The fertility of spikelets at high temperature can be used as a screening tool for high temperature tolerance during the reproductive phase. Selection for heat tolerance should be done for those breeding materials which can tolerate temperatures higher than 38°C (Satake and Yoshida, 1978). Some cultivars (N22) that are not so sensitive to relatively higher temperatures, have already been identified. Breeding for high-temperature stress focuses the use of these materials as genetic donors. For efficient selection in a breeding programme, visible markers of high temperature tolerance are required. In this regard, the length of the anther and the size of its basal pore are some of the morphological traits that can be easily identified and can be used in breeding and as screening tool for selection of rice germplasms for high-temperature resistance. Genetic modification of the male reproductive organ should be targeted in future breeding programmes as it is more sensitive to high temperature. Exploiting the existing genotypic variation in flowering time serves as a useful mitigation option for rising temperature as it is a relatively simple trait that needs to be focused in breeding programmes.

Several recent studies have used molecular markers to map the quantitative trait loci (QTL) underlying tolerance to HTS in rice. Two major QTLs controlling grain yield and spikelet fertility under high temperature stress has been identified on chromosome 1 (qHTSF1.1) and chromosome 4 (qHTSF4.1). These two major QTL could explain 12.6 per cent (qHTSF1.1) and 17.6 per cent (qHTSF4.1) of genetic variation for spikelet fertility under high temperature. The plants with qHTSF4.1 showed significantly higher spikelet fertility than other genotypes (Ye *et al.*, 2012). Similarly, Xiao *et al.*, 2010 have identified the same two major QTLs on chromosome 4 and chromosome 10 that affected the seed set percentage in rice.QTL mapping, along with associated genetic studies focusing on the relationship between the phenotype of a trait and its genetic markers, provide an opportunity to relate specific alleles to trait variation and thus to identify candidate genes (Cardon and Bell, 2001). The production of high-temperature tolerant transgenic rice cultivars would provide a stability advantage and will improve its overall performance under temperature stress.

Physiological and Molecular Responses of Rice Plants to Cold Stress

Low temperature stress is one of the environmental stresses affecting rice growth and development. Understanding the physiological mechanisms through which rice plants respond to low temperature is of fundamental importance to rice improvement programmes. Germination and seedling establishment are sensitive growth stages for rice to cold stress. Even though temperature does not prevent rice germination, it delays plant emergence. Optimum temperature range for rice germination lies between 20 and 35°C, and the temperature of 10°C is cited as the minimum critical value below which rice does not germinate (Yoshida,1980). Cold stress causes seedling mortality, spikelet sterility and eventually significant yield losses (Shimono *et al.*,2002) K Exposure to low non- lethal temperature usually induct a variety of biochemical, physiological and enzymatic changes in plant, which can result in an acclimation response that is characterized by a greater ability to resist injury or survive an otherwise lethal low temperature stress (Hughes *et al.*, 1996)

Mechanisms Underlying Low Temperature Stress Tolerance in Rice

Cold tolerance in rice is a very complex trait (Maruyama *et al.*, 2014). Cold stress affects chlorophyll content and fluorescence, and thus interferes with photosynthesis in rice (Kanneganti and Gupta 2008; Kim *et al.*, 2009). Moreover, increased contents of reactive oxygen species (ROS) and malondialdehyde (MDA) that accumulate during cold stress in rice can impair metabolism via cellular oxidative damage (Xie *et al.*, 2009).

Rice also possesses strategies to cope with or adapt to cold stress. Cold-treated rice plants accumulate proline, an amino acid that stabilizes protein synthesis, and thereby maintains the optimal function of rice cells (Kandpal and Rao, 1985). Under cold stress, contents of antioxidant species also increase to scavenge ROS and protect rice plants against oxidative damage (Sato *et al.*, 2011). Such physiological changes that occur upon cold treatment of rice, whether mediators or symptoms of cold damage, can also be used as indicators to evaluate the cold tolerance of rice.

Chlorophyll Content and Fluorescence Indicate Effects of Cold Stress on Photosynthesis

Cold stress can inhibit chlorophyll synthesis and chloroplast formation in rice leaves. Thus, reduced chlorophyll content can indicate the effect of low temperature on rice plants (Sharma *et al.*, 2005). During cold stress, Fv/Fm values decrease slightly in plants that tolerate cold, but decrease significantly in plants that are sensitive to cold (Bonnecarrère *et al.*, 2011).

Changes in Membrane Fluidity Initiate Cellular Cold Responsive

Changes in ambient temperature can affect cell membranes quickly, although this is a reversible process. Changes in membrane fluidity are measured by the membrane polarization index, p, an inverse indicator of membrane fluidity. Rice cells can sense cold stress based on changes in membrane rigidity, the physical state of membrane proteins, and osmotic pressure. Low temperatures initiate increased membrane rigidity.

ROS and MDA Mediate Cold Damage and Cold Sensing in Rice

In chloroplasts, ROS may cause over-reduction of the electron transport chain, limit CO_2 fixation, and interfere with the photosynthetic process. ROS can also cause damage during cold stress through their effects upon the electron transport chain in mitochondria (Suzuki and Mittler, 2006). ROS degrade polyunsaturated lipids to form MDA, a reactive aldehyde that initiates toxic stress in cells and subsequently causes cellular dysfunction and tissue damage (Pamplona, 2011). Accumulation of ROS in cells of cold-treated rice triggers expression of cold-responsive genes and regulation of the cold-responsive signaling network.

Soluble Sugars, Proline and Antioxidants Protect Rice from Further Damage Due to Cold Stress

Soluble sugars like sucrose, trehalose, raffinose, and stachyose contents can increase under low temperature, these metabolites can be used as indicators to evaluate the potential cold tolerance of rice varieties. Proline accumulation is also enhanced by cold stress. In addition to acting as a reservoir of carbon and nitrogen, proline also protects cellular enzymes from denaturation. Increased proline content has been widely observed in rice varieties under low temperatures. Finally, the significant correlations between proline contents and cold tolerance have help to confirm the function of proline during the cold response in rice (Kim and Tai, 2011).

Recent Strategies for Breeding Low Temperature Stress Tolerance in Rice

QTL identified in various cultivars facilitate the breeding of cold-tolerant rice. Cold tolerance in rice is a quantitative trait controlled by multiple genes. Because it is often difficult to directly associate plant phenotypes with the genes responsible for cold tolerance, marker-assisted selection is an effective means of developing cold-tolerant cultivars (Shirasawa *et al.*, 2012). The development of molecular markers and linkage maps has made it possible to identify QTL that control cold tolerance in rice. QTL analyses have been carried out using rice populations with large levels of genetic variation for cold tolerance. Two major QTL (qCTP11 and qCTP12) for cold tolerance at the plumule stage were identified. QTL, qCTS12a, on chromosome 12 accounted for 41 per cent of the phenotypic variation in seedling growth after cold stress. Two QTLs, Ctb1 and Ctb2, on chromosome 4 were found to confer cold tolerance at the booting stage of rice.

Conclusion and Future Outlook

Climate change is likely to have an adverse impact on rice production in India. This is mainly due to extreme weather events such as higher temperatures and sea-level rise such as drought, submergence and salinity. Progress on developing rice varieties for the unfavorable areas has shown that modern breeding tools can address many problems of farmers in these areas. Future progress may be more rapid if we can effectively use advances in genomics and recent biotechnological innovations. Achieving impacts in these areas will require continued investment in crop improvement research as well as improvements in delivery of improved seeds to the farmers. This provides some optimism that the adverse effects of climate change in rice-growing areas can be partially compensated. In conclusion, to feed the ever growing population, we need to solve the abiotic stress problems in rice and this is the principal challenge for Plant Physiologist, Plant Breeders and Plant biotechnologist.

References

Azhiri-Sigari, T., A.Yamauchi,A. Kamoshita and L. Wade. 2000. Genotypic variation in response ofrainfed lowland rice to drought and rewatering II. Root growth. *Plant Production Science* 3, 180-188.

Bao, F., J.J.Shen Brady S.R, G.K. Muday,T. Asami, and Z.B. Yang. 2004. Brassinosteroids interact withauxin to promote lateral root development in Arabidopsis. *Plant Physiology* 134, 1624-1631.

Bernier J., A. Kumar, V.Ramaiah, D. Spaner, G. Atlin. 2007. A large-effect QTL for grain yield under reproductive-stage drought stress in upland rice. *Crop Sci.*47: 507-518.

Blumwald, E., and A. Grover. 2006. Salt tolerance. In: Halford NG (ed) Plant biotechnology: current and future uses of genetically modified crops. John Wiley and Sons Ltd., UK. 206–224.

Bonnecarrère V, Borsani O, Díaz P, Capdevielle F, Blanco P, Monza J.2011. Response to photoxidative stress induced by cold in japonica rice is genotype dependent. *Plant Sci* 180(5): 726–732

Cardon, L.R. and J.I. Bell. 2001. Association study designs for complex diseases. *Nature Reviews Genetics* 2, 91–99.

Fukuda A., A. Nakamura, A. Tagiri, H. Tanaka, A. Miyao, H. Hirochika andY. Tanaka. 2004. Function, intracellular localization and the importance in salt tolerance of a vacuolar $Na^{(+)}/H^{(+)}$ antiporter from rice. *Plant Cell Physiol.*, 45: 146–159.

Garcia A.B., J.E. de Almeida, S. Iyer, T. Gerats M. Van Montagu, and A.B. Caplan. 1997. Effects of osmoprotectants upon NaCl stress in rice. *Plant Physiol.* 115: 159–169.

Ge, L., H. Chen, J.F. Jiang, Y. Zhao,M.L. Xu, Y.Y. Xu, Tan, K.H, Z.H. Xu and K. Chong. 2004. Overexpression of OsRAA1 causes pleiotropic phenotypes in transgenic rice plants, including altered leaf, flower, and root development and root response to gravity. *Plant Physiology* 135, 1502-1513.

Hughes, J.C.,Ayers, J. E. and Swain, T. 1996. After-cooking blackening in potatoes. I. Introduction and analytical methods. *J. Sci.Food and Agri.*, 13, 224–229.

Kandpal R.P., Rao, N.A. 1985. Alterations in the biosynthesis of proteins and nucleic acids in finger millet (Eleucine coracana) seedlings during water stress and the effect of proline on protein biosynthesis. *Plant Sci* 40(2): 73–79

Kanneganti V, Gupta A.K. 2008. Overexpression of OsiSAP8, a member of stress associated protein (SAP) gene family of rice confers tolerance to salt, drought and cold stress in transgenic tobacco and rice. *Plant Mol Biol* 66(5): 445–462

Kawasaki, S., C. Borchert, M. Deyholos, H. Wang, S. Brazille, K. Kawai, D. Galbraith and H.J. Bohnert. 2001. Gene expression profiles during the initial phase of salt stress in rice. *Plant Cell.*, 13: 889–905

Khan M.S.A., A. Hamid andM.A. Karim. 1997. Effect of sodium chloride on germination and seedling characters of different types of rice (*Oryza sativa* L.). *J Agron Crop Sci.*, 179: 163–169.

Kim S-I, Tai, T.H. 2011. Evaluation of seedling cold tolerance in rice cultivars: a comparison of visual ratings and quantitative indicators of physiological changes. *Euphytica* 178(3): 437–447

Koyama, M.L., A. Levesley, R.M. Koebner, T.J. Flowers and A.R. Yeo. 2001. Quantitative trait loci for component physiological traits determining salt tolerance in rice. *Plant Physiol.*, 125: 406–422.

Larkindale, J. and E. Vierling. 2008. Core genome responses involved in acclimation to high temperature. *Plant Physiology.* 146, 748–761.

Li Zi, Y. and Chen Shou, Y. 1999. Inducible expression of translation elongation factor 1A gene in riceseedlings in response to environmental stresses. *Acta Botanica Sinica.Aug., 1999;* 41, 800-806.

Lilley, J.M and M.M. Ludlow. 1996. Expression of osmotic adjustment and dehydration tolerance indiverse rice lines. *Field Crops Research* 48, 185-197.

Lin H.X., M.Z. Zhu, M. Yano, J.P. Gao, Z.W. Liang, W.A. Su, X.H. Hu, Z.H. Ren and D.Y. Chao. 2004. QTLs for Na+ and K+ uptake of the shoots and roots controlling rice salt tolerance. *Theoretical and AppliedGenetics.*108, 253-260.

Liu L., H.R. Lafitte and D. Guan. 2004b. Wild *Oryza* species as potential sources of drought- adaptive traits. *Euphytica* in press.

Lutts S., J.M. Kinet, J. Bouharmont. 1995. Changes in plant response to NaCl during development of rice (*Oryza sativa* L.) varieties differing in salinity resistance. *J Exp Bot.* 46: 1843–1852.

Maestri, E., N. Klueva, C. Perrota,M. Gulli, H.T. Nguyen andN. Marmiroli. 2002. Molecular genetics of heat tolerance and heat shock proteins in cereals. *Plant Molecular Biology.* 48, 667–681.

Maruyama K, Urano K, Yoshiwara K, Morishita Y, Sakurai N, Suzuki H, Kojima M, Sakakibara H, Shibata D, Saito K. 2014. Integrated analysis of the effects of cold and dehydration on rice metabolites, phytohormones, and gene transcripts. *Plant Physiol* 164(4): 1759–1771.

Matsui, T. and K. Omasa. 2002. Rice (*Oryza sativa* L.) cultivars tolerant to high temperature at flowering: anther characteristics. *Annals of Botany*. 89, 683–687.

Matsui, T. and H. Kagata 2003. Characteristics of floral organs related to reliable self pollination in rice (Oryza sativa L.). *Annals of Botany*. 91, 473–477.

Matsui, T., K. Omasa andT. Horie. 1999. Mechanism of anther dehiscence in rice (*Oryza sativa* L.). *Annals of Botany*. 84, 501–506.

McCouch, S., L. Teytelman,Y. Xu,K. Lobos,K. Clare,M. Walton,B. Fu, R. Maghirang, Z. Li,Y. Xing, Q. Zhang,I. Kono,M. Yano, R. Fjellstrom, G. DeClerck, D. Schneider, S. Cartinhour, D. Ware and L. Stein. 2003. Development and mapping of 2240 new SSR markers for rice (*Oryza sativa* L.). *DNA Research* 9, 199-207.

Moons A., G. Bauw, E. Prinsen, M. Van-Montagu and V.D.D. Straeten. 1995. Molecular and physiological responses to abscisic acid and salt in roots in salt-sensitive and salt tolerant indica rice varieties. *Plant Physio.,l* 107: 177–186.

Nakagawa, H., T. Horie and T. Matsui. 2003. Effects of climate change on rice production and adaptive technologies. In Rice Science: Innovations and Impact for Livelihood. Proceedings of the International Rice Research Conference, Beijing, China, 16–19 September 2002 (Eds T.W. Mew, D. S. Brar, S. Peng, D. Dawe and B. Hardy), pp. 635–658. Manila, The Philippines: IRRI.

Nguyen, H.T., R.C. Babu and A. Blum. 1997. Breeding for drought resistance in rice: Physiology andmolecular genetic considerations. *Crop Science* 37, 1426-1434.

O'Toole, J.C.1982. Adaptation of rice to drought-prone environments. In '*in* Drought Resistance in Crops, with Emphasis on Rice'. 195-213. (International Rice Research Institute, P.O. Box 933, Manila, Philippines).

Pamplona, R. 2011. Advanced lipoxidation end-products. *Chem-Biol Interact* 192 (1): 14–20

Pareek A, S.L.,Singla andA. Grover. 1997. Salt responsive proteins/genes in crop plants. In: Jaiwal PK, Singh RP, Gulati A (eds) Strategies for improving salt tolerance in higher plants. Science Publishers, USA. 365–382.

Satake, T. andS. Yoshida. 1978. High temperature-induced sterility in indica rices at flowering. *Japanese Journal of Crop Science* 47, 6–17.

Sato Y, Masuta Y, Saito K, Murayama S, Ozawa K. 2011. Enhanced chilling tolerance at the booting stage in rice by transgenic overexpression of the ascorbate peroxidase gene, OsAPXa. *Plant Cell Rep* 30(3): 399–406

Setter,T.L. and Laureles, E.V. 1996. The beneficial effect of reduced elongation growth on submergence tolerance of rice. *J. Exp. Bot.*, **47:** 1551–1559.

Sharma P, Sharma N, Deswal R. 2005. The molecular biology of the low temperature response in plants. *BioEssays* 27(10): 1048–1059.

Shimono, H., Hasegawa, T. and Iwama, K. 2002. Response of growth and grain yield in paddy rice to cool water at different growth stages. *Field crop research*, 73: 67-79.

Shirasawa S, Endo T, Nakagomi K, Yamaguchi M, Nishio T.2012. Delimitation of a QTL region controlling cold tolerance at booting stage of a cultivar, 'Lijiangxintuanheigu', in rice, *Oryza sativa* L. *Theor Appl Genet* 124: 937–946

Sivakumar P., P. Sharmila andP.P. Saradhi. 1998. Proline suppresses Rubisco activity in higher plants. *Biochem Biophys Res. Commun.*, 252: 428–432.

Song, Z. P., B.R. Lu and K.J. Chen. 2001. A study of pollen viability and longevity in Oryza rufipogon, O. sativa and their hybrids. *International Rice Research Notes* 26, 31–32.

Stiller, V andJ. Lafitte RandSperry.2003. Hydraulic properties of rice and the response of gas exchange to water stress. *Plant Physiology* 132, 1698-1706.

Suzuki N, Mittler R. 2006. Reactive oxygen species and temperature stresses: a delicate balance between signaling and destruction. *Physiol Plant* 126(1): 45–51

Tilman D, Balzer C, Hill J, Befort BL (2011) Global food demand and the sustainable intensification of agriculture. *Proc Natl Acad Sci U S A* 108: 20260-20264.

Venuprasad R., C.O. Dalid, M. Del Valle, D. Zhao, M. Espiritu, M.T.S. Cruz, M. Amante, A. Kumar and G.N. Atlin. 2009. Identification and characterization of large-effect quantitative trait loci for grain yield under lowland drought stress in rice using bulk- segregant analysis. *Theor. Appl. Genet.*120: 177-190.

Verulkar S.B., N.P. Mandal, J.L. Dwivedi, B.N. Singh, P.K. Sinha, R.N. Mahato, P. Dongre, O.N. Singh, L.K. Bose, P. Swain, S. Robin, R. Chandrababu, S. Senthil, A. Jain, H.E. Shashidhar, S. Hittalmani, C. Vera Cruz, T. Paris, A. Raman, S. Haefele, R. Serraj andG. Atlin and A. Kumar. 2010. Breeding resilient and productive genotypes adapted to drought-prone rainfed ecosystem of India. *Field Crop Res.* 117: 197-208.

Wahid, A., S. Gelani, M. Ashraf and M.R. Foolad. 2007. Heat tolerance in plants: an overview. *Environmental and Experimental Botany* 61, 199–223.

Xiao, Y., Y. Pan, L. Luo, G. Zhang, H. Deng, L. Dai, X. Liu, W. Tang, L. Chen, and G. Wang, 2011: Quantitative trait loci associated with seed set under high temperature stress at the flowering stage in rice. *Euphytica* 178, 331-338.

Xie G, Kato H, Sasaki K, Imai R. 2009. A cold-induced thioredoxin h of rice, OsTrx23, negatively regulates kinase activities of OsMPK3 and OsMPK6 *in vitro*. *FEBS Lett* 583(17): 2734–2738

Ye, Changrong., Argayoso, May A., Redoña, Edilberto D., Sierra, Sheryl N., Laza, Marcelino A., Dilla, Christine J., Mo, Youngjun., Thomson, Michael J., Chin, Joonghyoun., Delaviña, Celia B., Diaz, Genaleen Q., Hernandez, Jose E. 2012. Mapping QTL for heat tolerance at flowering stage in rice using SNP markers. *Plant Breeding* Vol. 131 issue 1 February 2012. pp. 33-41.

Yeo A.R, S.J.M. Caporn and T.J Flowers. 1985. The effect of salinity upon photosynthesis in rice (*Oryza sativa* L.): gas exchange by individual leaves in relation to their salt content. *J. Exp. Bot.* 36: 1240–1248.

Yin, X., M.J. Kroff, and J. Goudriann. 1996. Differential effects of day and night temperature on development to flowering in rice. *Annals of Botany.* 77, 203–213.

Yoshida, S. 1980. Fundamentals of rice crop science. pp. 1-110. IRRI, Los Banos, Philippines.

Chapter 14

Seed Production Techniques in Varieties and Hybrids of Rice

M. Bhaskaran

Special Officer (Seeds)
Seed Centre, TNAU,
Coimbatore, Tamil Nadu
e-mail: bhaskm@yahoo.co.in

The growing population of India along with its changing food habits needs around 130 million tonnes of rice by 2025. Seed is the basic and critical input in crop production. The usage of farm saved seed is to be reduced and farmers are to be encouraged to use quality seed to increase productivity levels. The increase in rice production can be achieved through quality seed, hybrid seed and transgenic varieties tolerant to pests etc. But the production in India is low when compared with other countries particularly to China which has the second largest area under rice. To meet the rice growth central and state governments generated many research and extension programme in India. When the rice varietal improvement programme has not achieved a substantial improvement in yield, exploitation of heterosis through the development of rice hybrids has been identified as an alternate strategy to break the yield plateau in rice productivity because China has recorded tremendous increase in rice yield through hybrid rice technology. The average yield in China is 6.7 t/ha. The father of hybrid rice *Yuvan Long Ping* developed a hybrid rice which yielded 17 t ha^{-1} in china.

In India, work on hybrid rice was started in 1989 by ICAR and was further strengthened with the assistance from UNDP/FAO since 1991. The first hybrid CORH 1 (MGR) was released in 1994 at TNAU, Coimbatore and so far more than 50 hybrids have been released from public sector.

Botany

Paddy is a self-pollinated crop with cross-pollination upto 0-4 per cent. The flower opening starts from the tip of the primary and secondary branches and proceeds downwards (basipetal). Normally 6-8 days are required to complete flowering in a panicle. Under normal condition flower opening is between 7 to 10 a.m. The flower remains open for 10 minutes and afterwards it closes. The dehiscence of anthers is independent of spikelet opening. The dehiscence may take place before opening up of flowers or after flower opening. The stigma is receptive for 3 days and pollen grains viable for 15 -20 minutes.

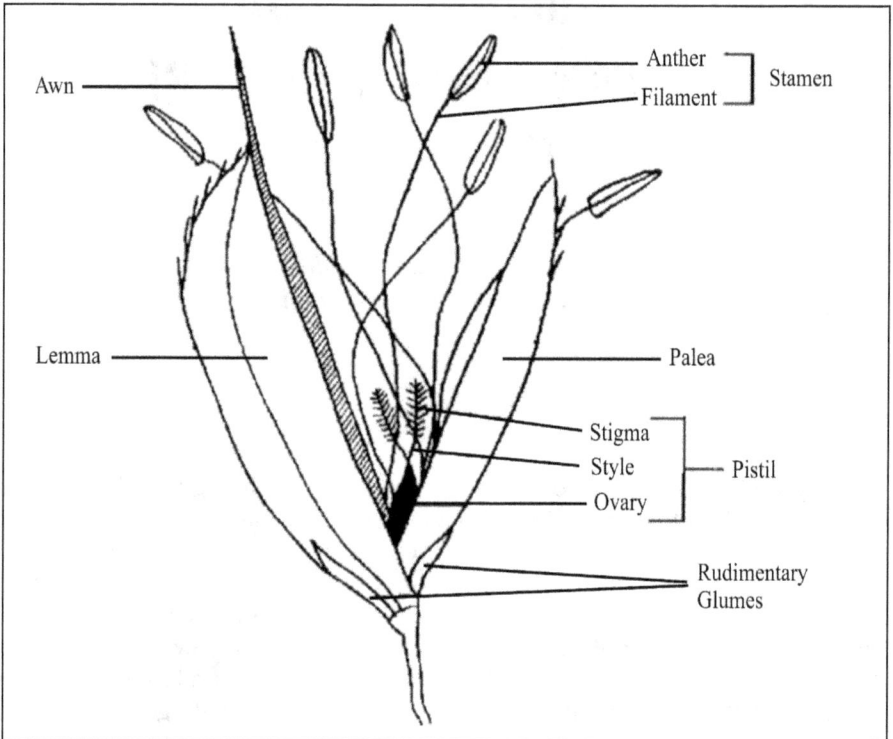

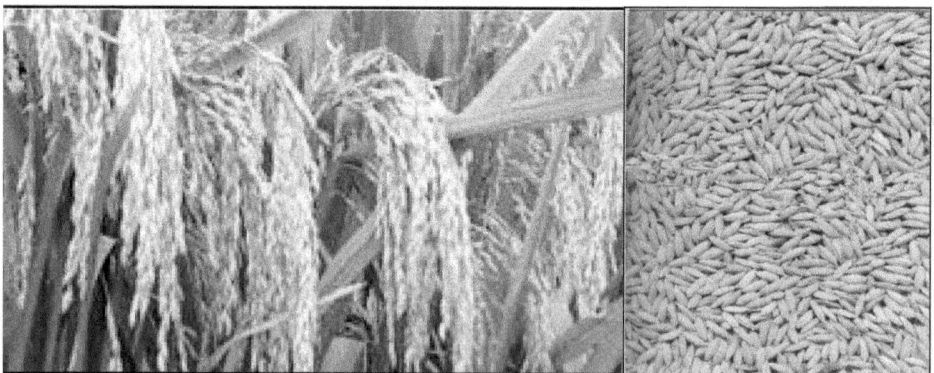

Figure 14.1: Rice Inflorescence.

Methods of Seed Production

a. Varieties

Nucleus seeds are raised in ear to row method under isolation. The rows containing off types are removed. True to type earheads are marked and harvested separately and issued for further multiplication. In varietal seed multiplication, seeds are allowed to set by self pollination that are raised in isolation.

b. Hybrids

The tools involved in hybrid seed production are cytoplasmic genic male sterility (CGMS) and environmental genetic male sterility (EGMS) and single line breeding (Apomixis) systems. In India the common method for hybrid seed production is CGMS otherwise called as three line breeding system, where three lines (A, B and R lines) are involved. A line is a male sterile line and serves as female parent of F_1 hybrid. B line is the maintainer line of A line and is male fertile. It is isogenic to A line in all aspects except male fertility. R line is the male line in actual hybrid seed production. It restores the fertility of A line and hence it is known as restorer line.

Stages of Multiplication

Variety

In varietal seed production, the three tier seed multiplication system is adopted. The nucleus seed is multiplied as breeder, foundation and certified seed with the maintenance of genetic purity as above 99 per cent. But on high demand for seed, since the crop is self pollinated four tier system of seed multiplication as breeder seed, foundation seed I, foundation seed II and certified seed is adopted.

Hybrid

In hybrid seed production programme particularly in breeder and foundation seed stages, A line is multiplied with the use of B line and is produced in isolation from R line which is multiplied as that of variety. In certified seed production A line and R line are crossed to produce F1hybrid seed.

Land Requirement

Land should be fertile with good irrigation and drainage facilities. It should be free from volunteer plants. It should have good sunlight and aeration. The seed crop should be isolated from other varieties. The field should not have been grown with the same variety/hybrid in the previous season. In varietal seed production alone if it is the same class of seed for the same variety and approved by seed certification agency, then it is accepted for seed production.

Isolation

Isolation distance is 3 m for varieties in both foundation and certified class of seed. For hybrid, the isolation requirement is 200 and 100 m for foundation and certified seed stages, respectively. When space isolation is not possible, the time isolation of over 21 days or barrier isolation with polythene sheets of 2m height or

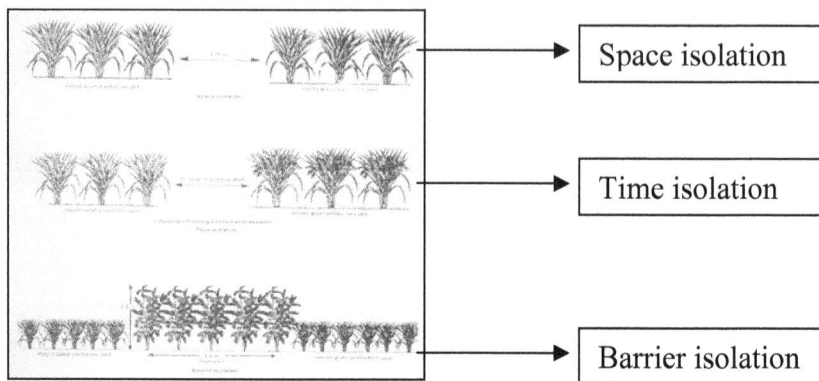

Figure 14.2: Isolation Methods for Hybrid Rice Production.

barrier crops like sesbania, sugarcane and maize covering a distance of 3m would serve the purpose of isolation meant for restricting outcrossing for the maintenance of genetic purity.

Season

Varieties

Seasons for production of rice of Tamil Nadu vary with area and duration as follows.

Season	Month of Sowing	Duration of the Varieties
Navarai	December - January	Below 120 days
Sornavari, Early kar	April - May	Below 120 days
Kar	May - June	Below 120 days
Kuruvai	June - July	Below 120 days
Early samba	July- August	130 - 135 days
Samba	August	130-135 and above 150 days
Late samba/Thaladi/Pishanam	September - October	130-135 days
Late thaladi	October-November	115 - 120 days

Hybrids

Seed set and seed yield will be affected if temperature is below 20°C and above 35°C during the time of flowering. The RH should range from 70-80 per cent. The difference in temperature between day and night should be 8-10° C. There should be sufficient sunshine with moderate wind velocity (2-3m/sec.). In Tamil Nadu ideal time for sowing during Kharif is 2nd fortnight of May and during Rabi 2nd fortnight of December. Between the season Rabi is ideal and more suitable than Kharif.

Popular Varieties and Hybrids of Tamil Nadu

Varieties

Crop	Popular Varieties in Seed Supply Chain	New Varieties
RICE	**State Varieties :** ADT 36, ADT 37, ADT 38, ADT 39, ADT 42, ADT 43, ADT 44, ADT (R) 45, ADT (R) 46, ADT (R) 47, ADT (R) 48, ASD 16, ASD 18, ASD 19, CO 43, CO 47, CO (R) 48, CO (R) 49, CO (R) 50, CO (R) 51, CR 1009, I.W.Ponni, TRY 1, TPS 3, TKM 9, Paiyur 1, PMK (R) 3, Anna (R) 4	ADT (R) 48, ADT (R) 49, ADT (R) 50, Anna (R) 4, CO (R) 49, CO (R) 50, CO (R) 51*
	National Variety : BPT 5204, NLR 34449, CR 1009, IR 36, IR 50	

Hybrids

Hybrid	Year of Release	Duration in Days	Parental Lines
CORH 1 (MGR)	1994	110-115	IR 62829 A x 1098-662 R
CORH 2	1998	120-125	IR 58025 A x C 20 R
ADTRH 1	1998	110-115	IR 58025 A x IR 66 R
CORH 3	2006	110-115	TNAU CMS 2A X CB 87 R
TNAU Rice Hybrid CO 4	2006	110 -115	TNAU COMS 2A X CB 87 R

Seeds and Sowing

The basic seeds should be obtained from authenticated seed source with respective class certification seed tag and purchase bill. The seed rate for varieties is based on their duration and it varies as 60 kg ha^{-1} for short duration varieties, 40 kg ha^{-1} for medium duration varieties and 30 kg ha^{-1} and for long duration varieties. In hybrid seed production, the seed requirement will be 20, 10 and 10 kg ha^{-1} for A, B and R lines, respectively. The seeds are sown in nursery beds and are transplanted in the main field.

Seed Selection and Treatment

Normally to protect the plant from leaf blast and other diseases the seeds are soaked with Bavistin 50WP @ 2 glit^{-1} for 24h against seed borne diseases. (2.5 g of chemical is required for 1 kg seed). The treatment can be given in a rotating drum or putting the seed in a container and shaking it well after adding chemical. The seeds also may be treated with *Pseudomonas fluorescens* @ 10g kg^{-1} of seed as dry treatment otherwise *P. fluorescens* will be soaked in water @ 10g kg^{-1} l practice it, then seeds are soaked in the solution @ 1: 1 for 16h, then incubated and sown in nursery. But as per requirement several presowing seed treatments are practiced as follows.

Pre-sowing Seed Management

Dormant Seed

In dormant cultivars, seeds are soaked in equal volume of 0.1 N conc. HNO_3 or in 0.5 per cent KNO_3 for a duration of 12-16 h to break the dormancy. The seeds are then dried to original moisture content.

Seed Quality Upgradation

By removal of ill filled and immatured seed, the seeds could be upgraded using specific gravity. For upgrading, the seeds are soaked in salt water (1.5 kg of common salt in 1 lit of water grading) raised to the specific gravity of 1.03, easily identified through the flotation of egg. The floaters are removed as illfilled and immatured, while sinkers are washed with adequate water and used for nursery sowing.

1.5 kg of salt dissolved in 10 lts. of water.
Partial floating of egg
(size of 25 paise coin above solution)

Separation of seed

Figure 14.3: Egg Floatation Technique for Paddy Seed Upgradation.

For Rainfed Rice or Direct Sowing

The seeds are hardened by soaking the seeds in equal volume of 1 per cent KCl (potassium chloride) solution for 16 h and are dried back to its original moisture content or seeds are primed with 6 per cent Pink Pigmented Facultative Methylotrophs (PPFMs) for 18 h.

Designer Seed for Direct Sowing and for Wet Nursery

This technique is being practiced to achieve the integrated benefit of all seed treatment. In this integrated seed management technique, seeds are hardened with 1 per cent KCl for 16 h and are dried back to their original moisture content and are coated with polymer @ 3 mlkg^{-1} + imidachloprid @ 2ml kg^{-1} + *Pseudomonas fluorescens* @ 10 g kg^{-1} + Azophos @ 120g kg^{-1} of seed.

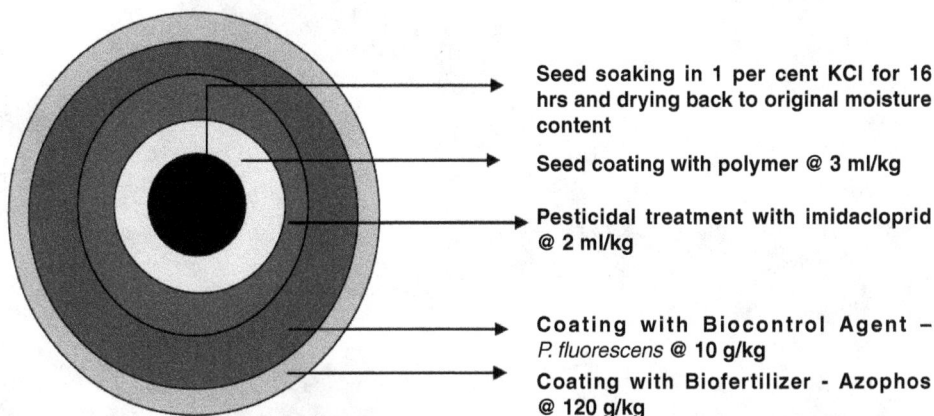

Seed soaking in 1 per cent KCl for 16 hrs and drying back to original moisture content

Seed coating with polymer @ 3 ml/kg

Pesticidal treatment with imidacloprid @ 2 ml/kg

Coating with Biocontrol Agent – *P. fluorescens* @ 10 g/kg

Coating with Biofertilizer - Azophos @ 120 g/kg

Figure 14.4: Designer Seed/Integrated Seed Treatment Technique of Rice.

Organic Seed Treatment

In organic farming seeds are primed with natural products. Seeds are soaked in 3 per cent cowpea sprout extract for 16 h in the seed to solution ratio of 1: 1 and are dried back to its original moisture content (or) the seeds are soaked in 80 per cent *Pseudomonas fluorescens* (80 g of powder form mixed in 100 ml of water) solution for 16h in adopting the seed to solution ratio of 1: 1.

Preparation of Pulse Sprout Extract

Cowpea seeds were soaked overnight and incubated in a wet cloth for 12 h to enable sprouting. Later, 100 g of sprouts were ground in a mixer grinder by using ice cubes of 100 ml of water to prepare extracts of 100 per cent concentration. The ground material was squeezed through cloth bag to extract the sprout extract (Figures 14.5–14.7).

Preparation of Seed Bed

For transplanting one hectare of land by conventional practice, a nursery area of 20 cents (800 m^2) would be sufficient. The area under nursery sowing can be further lowered down to 100 sq.m by adopting System of Rice Intensification (SRI) and Integrated Crop Management (ICM) method of rice cultivation.

Figure 14.5: Cow Pea Sprouting.

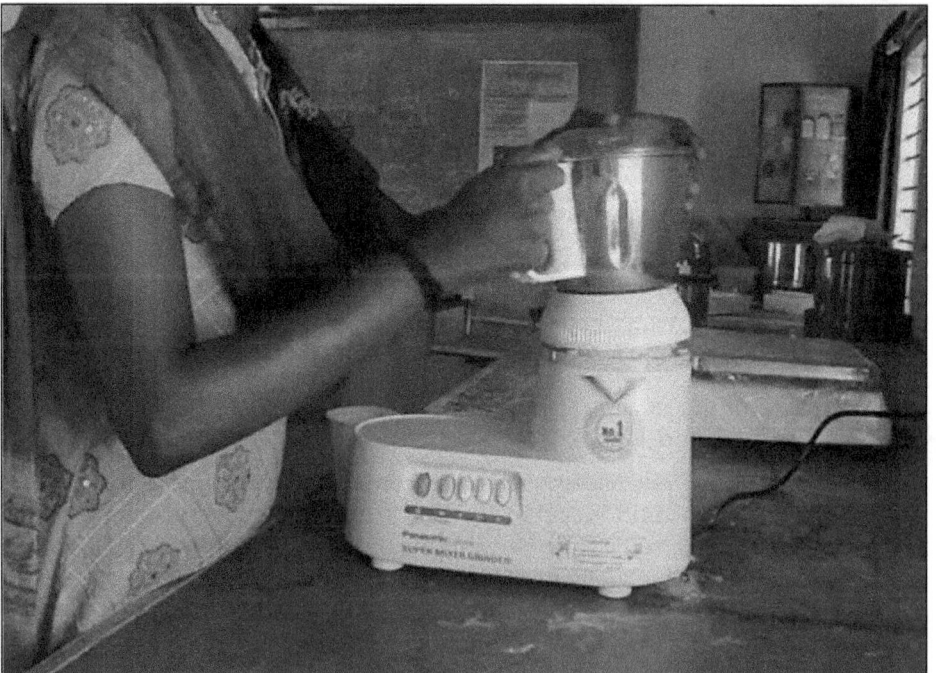

Figure 14.6: Grinding of Sprouts with Cold Water.

Figure 14.7: Squeezing.

Raised Seed Bed

Normally a raised bed is preferred for raising seedlings. A bed of 10 cm height from ground level in high rainfall areas is necessary. Size of each bed may be 10 m in length and 1.25 m width with 30 cm wide channel in between the two seed beds. Treated seeds should be evenly broadcasted in each bed after manuring.

Wet Nursery

Nursery should be thoroughly ploughed and perfectly leveled so that a thin layer of water is maintained during the emergence. After the manuring and puddling, sprouted seeds are uniformly broadcasted in each seed bed. To get sprouted seeds, overnight treated and soaked rice seeds filled in wet gunny bags are kept for about 48 h.

Nursery Management

In nursery, on the occasion of raising nurseries of different genotypes nearby, separate irrigation channels should be formed for each genotype.

In hybrid seed production female and male nurseries should be raised separately. Sparse sowing in nursery beds @ 1 kg^{-1} cent should be practiced to obtain robust seedlings. Application of DAP @ 2 kg^{-1} per cent, if not possible apply straight fertilizers 16 kg of urea and 120 kg of super phosphate is recommended for production of elite seedling in nursery. Basal application of DAP is also recommended when the seedlings are to be pulled out in 20 - 25 days after sowing.

If seedlings are to be pulled out at 25 days, application of DAP is to be done 10 days prior to pulling out. In case of clayey soils where root snapping is a problem, DAP has to be applied @ 1 kg per cent at 10 days after sowing.

Advantages of Phosphorus Application to Nursery

1. Seedlings absorb and store phosphorous and utilize even at later stages of crop growth.

2. If 30 per cent recommended phosphorus as per soil test is applied to main field besides nursery application, higher yield can be realised.

3. Application of phosphorus to nursery is very economical.

Figure 14.8: Nursery Sowing.

Figure 14.9: Nursery Weeding.

Age of Seedlings

The optimum age of seedlings for transplanting is 18-22 days for short, 25-30 days for medium and 30-35 days for long duration varieties. In hybrids, seedlings of 20 - 25 days old have to be planted (hybrid) but as per synchronization requirement it varies with parental lines. For proper synchronization of flowering of male and female parents in hybrid seed production, staggered sowing should be done, which differ with hybrids as follows.

Hybrid	Season	Staggering
CORH 1	May–June	Male parent (R line) should be sown 5 days and 10 days after the sowing of female
	Dec–Jan	Male parent (R line) should be sown 10 days and 15 days after the sowing of female
CORH 3	Dec–Jan	Restorer (R) line, the male parent should be sown 6 days (R1), 3 days (R2), earlier than A line sowing and same day (R3) on the sowing of male sterile (A) line.
	May–June	Restorer (R) line, the male parent should be sown 3 days (R1) earlier, same day (R2) on the sowing of male sterile (A) line, the female parent and 3 days later (R3) the sowing of male sterile (A) line, the female parent
CORH 4	Nov–Dec	Restorer (R) line, the male parent should be sown 5 days (R1), earlier than A line sowing and same day (R2) on the sowing of male sterile (A) line, the female parent

For A line seed production B line should be sown 6 and 10 days after the sowing of A line in both the season. The days of staggering varies according to the location, season and duration of parents. Seedlings are to be transplanted at the age of 25 days.

Pulling Out of the Seedlings

The seedlings are pulled out at the appropriate time and tie the seedlings into a convenient bundle of 5-8 cm diameter with soft materials such as banana twine and keep the root portion submerged in water. The seedlings should not be allowed to dry.

Root Soak Treatment

Before transplanting the seedlings are given root soak treatment with 100 ml of chlorpyriphos 20 EC and also nutrient dip with urea @ 2.5 kg per liter water for 20 minutes.

Main Field Preparation

The field should be puddled and leveled well for anchoring of seedlings. The pulled seedlings are transplanted at 5 cm water level adopting the required spacing.

Spacing

In varietal seed production for Short duration varieties a spacing of 20 x 10 cm for medium duration varieties 20 x 15 cm and for long duration varieties a spacing

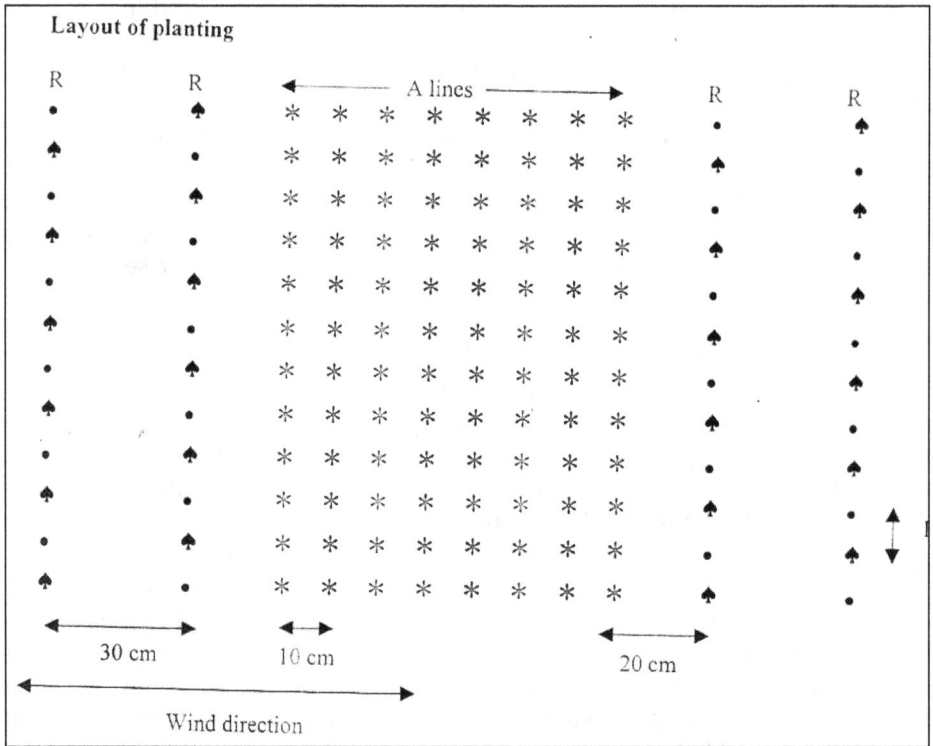

Figure 14.13: Row Ratio, Row Direction, Spacing and Planting Pattern for Hybrid Rice Seed Production.

Manures and Fertilizers

Farmyard manure can be applied on last puddling @ 12.5 tonnes/ha. For varieties the recommendation of NPK is as short duration : NPK @ 120: 40: 40 kg ha^{-1}, medium duration : NPK @ 150: 50: 60 kg ha^{-1} and for long duration : NPK @ 150: 50: 80 kg ha^{-1} and N is applied in three split doses. The recommended dose of NPK for the hybrid seed production is 150: 60: 60 kg ha^{-1}. The P is added at last puddling stage. The N and K are applied in three splits *viz.*, (1) Basal (2) Active tillering and (3) Panicle initiation stage. Additional nitrogen application delays panicle development whereas P and K promote the same.

Bio-fertilizer Application

Bio-fertilizers are applied through seed @ 3 packets ha^{-1} and to the main field @ 10 packets ha^{-1}. On application of bio-fertilizers only 75 per cent of recommended N has to be applied to the soil. For low land rice Azospirillum strain IPI responds well.

Application of Zinc Sulphate

Mix 25 kg of zinc sulphate with 50 kg of sand and apply the $ZnSO_4$ + sand mixture uniformly on the leveled field. Do not incorporate in the soils as there will be $ZnSO_4$ interaction. When green manure (6.25 t ha^{-1}) is applied, it is enough to apply 12.5 kg $ZnSO_4$ ha^{-1}.

Application of Lime

Apply lime to acid soils based on the soil analysis for obtaining normal rice yields. Apply 2.5 tonnes of lime ha^{-1} before last ploughing.

Basal Application of Gypsum

Gypsum at 500 kg/ha applied as basal with NPK in the main field in non calcareous heavy soils.

Weed Management

Mostly hand weeding is practiced. However, in the absence of manpower, herbicides like Butachlor 50 EC 1.25 a.i. ha^{-1} or commercial Machete @ 2.5 lit ha^{-1} in 3-4 cm standing water within 2-4 days after transplanting (DAT) is applied to control grassy weeds. or Pendimethalin (Stomp) @ 1.0 kg a.i. ha^{-1} in 400 litre water. To control sedges (*Cyperus* spp.) and broadleaved weeds, apply 2, 4-D 4 per cent granules @ 20 kg ha^{-1} at 4-5 days after transplanting. Rotary paddy weeder (Cono weeder or Japanese paddy weeder) can be used for weeding by running the weeder in between the rows. Use of cono weeder requires only about 4 to 5 labourers ha^{-1} compared to about 20 labourers/ha for hand weeding. It also helps in simultaneous incorporation of weed biomass into the soil and improves soil aeration and root respiration. For SRI and ICM practice, weeding should be done at 15 days interval up to maximum tillering stage. Line transplanting using a spacing of at least 20 cm between rows is prerequisite for using mechanical weeder. Alternatively, farmers can make their paddy weeder using a wooden plank and inserting removable pegs in the floats. At least two weedings (25 and 45 DAT) are required for higher productivity in low land rice.

Water Management

Rice is a water loving plant and requires about 3000 to 5000 litre of water to produce a kilogram of rice grain depending upon the system of cultivation. However, it is experimentally proved that by modifying management practices, adopting water efficient varieties, enhancing irrigation efficiency, reducing conveyance losses, etc., the water requirement of rice could be reduced substantially without reducing the productivity. Irrigation should be given during critical growth stages like tillering, panicle initiation, panicle emergence and grain filling stage to get maximum yield. Film of water *i.e.*, about 5cm of staggering water has to be maintained in the field throughout the crop period.

Flowering and Synchronization

The flowering period of male (6-8 days) and female (8-10 days) varies between the parents. Perfect synchronization of flowering between male and female parent is essential for seed set. The synchronization can be achieved by adopting any one of the following techniques.

Staggered Sowing

By sowing the male line (early parent) in different dates the flowering of both parents will coincide with female. In nursery sowing of early parent (male) can be 2-3 days later than late parent (female) which helps supply of pollen continuously for increased seed set.

Nitrogen Application

Apply Urea @ 35 kg ha^{-1} to the advancing parent (Flowering delayed due to enhancing of vegetative growth by application of urea) or spray (2-3 sprays) 20 kg ha^{-1} of urea in Knapsack sprayer in 500 lit. of water instead of power sprayer. This should be done from 4th stage of panicle initiation, which is around 70 days after sowing. The parents with delayed flowering can also be sprayed with DAP @ 2 per cent for effective synchronization.

Withholding of Irrigation

Draining of water in R line can delay its flowering by 2-3 days.

GA$_3$ Application

The panicle exertion in female parent is not full and incomplete. Good panicle exertion will help in improving the seed set. Hence, GA is sprayed especially in paddy. Spraying of GA$_3$ @ 75g ha^{-1} as 40 g of GA$_3$ at 5-10 per cent panicle emergence stage and the remaining 35 g of GA$_3$ at 24 h after first spray with knapsack sprayer at 500 litres of spray solution per hectare will increase the seed set and final yield. Morning 8 to 10 AM and evening 5 to 6 PM are ideal for spraying of GA$_3$.

> *Note: GA$_3$ is not soluble in water. Hence it should be dissolved with little amount of (1 g in 10 to 20 ml) 75 per cent alcohol and then volume is made up to the required quantity.*

Supplementary Pollination

Rope Pulling/Rod Driving

Rope pulling or shaking the pollen parent with the help of two bamboo sticks at 30-40 per cent of spike lets opening stage. This process is repeated for 3 – 4 times during the day time (10 am to 1 pm) at an interval of 30 min. This has to be repeated for 7 – 10 days during flowering period. Passing of rope or rod across the population 3 to 4 times daily for 7-10 days during anthesis will supplement the pollination mechanism and aid in out crossing in hybrid seed production. The normal anthesis time is between 10 AM - 1.00 PM and 3-4 PM.

Figure 14.14: Rope Pulling.

Figure 14.15: Rod Driving.

Rogueing

It is the unique feature of seed crop, that helps in maintenance of genetic purity of crop varieties. Rogueing is practiced from sowing upto harvest and plants differ

Figure 14.16: Rogueing.

Figure 14.17: Removal of Off-type.

Figure 14.18: Abnormality of Panicle.

Figure 14.19: Purple Tip.

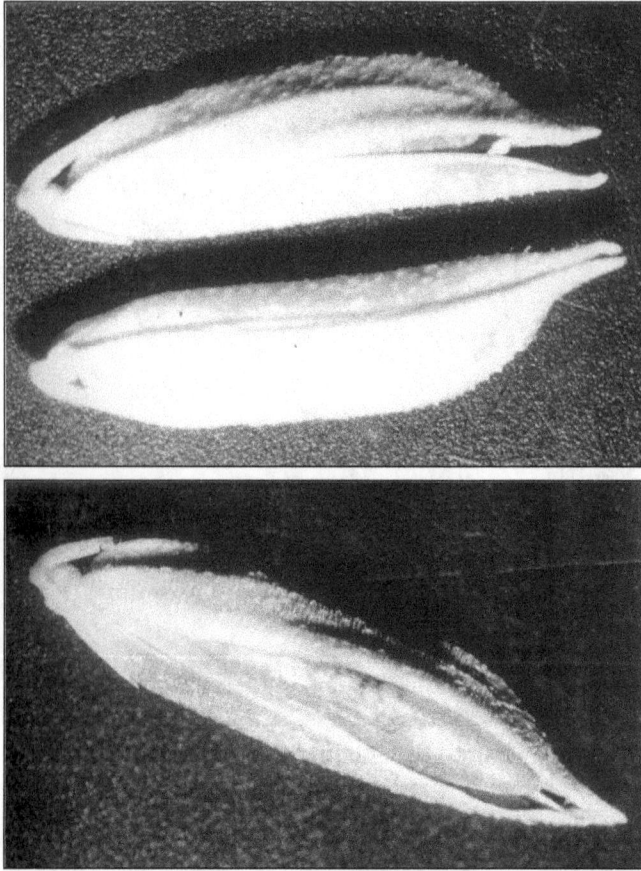

Figure 14.20: Split Husk.

in character as per release and notification are removed regularly. In paddy plant height, leaf colour, boot leaf angle, panicle type, its exertion, glume colour, seed size, kernal colour, petal spot are practiced for identification of off-types.

Figure 14.21: Wild Rice – Objectionable Weed.

From vegetative phase upto harvest the seed production plot should be checked for rogueing out volunteer, off type and diseased plants. Rogueing should be done daily from earhead emergence to maturity stage. The pollen shedders (presence of B line in A line) and other off types are to be checked at all times and the same should be removed to maintain genetic and physical purity of seeds. The accuracy of rogueing is checked for 2 times *i.e.*, before and after flowering by Seed Certification Officer.

Characters	Maximum Permitted (per cent)	
	Foundation Seed	Certified Seed
Varieties		
Off types	0.05	0.20
Objectionable weed plants (wild rice)	0.01	0.02
Hybrids		
Off types in seed parent	0.05	0.20
Off types in Pollinator	0.05	0.20
Pollen shedders in female	0.05	0.10
Objectionable weed plants	0.01	0.02

Foliar Spray

Foliar nutrition helps in maximizing the seed set. Spray with 2 per cent DAP (Diammonium phosphate) (or) 0.5 per cent Nutri Gold (organic growth promoter) (or) 3 per cent cowpea sprout extract (or) 3 per cent horse gram sprout extract at boot leaf stage and at 5-10 per cent flowering is recommended for both the varieties and hybrids. Foliar application with one per cent PPFMs (Pink Pigmented Facultative Methylotrophs) at book leaf initiation, flowering and milky stages is also recommended to increase the seed set and to resist the drought.

Plant Protection Measures

Integrated insect pest and disease management is the best option as it reduces farmers investment, utilizes farmers' knowledge and on-farm resources effectively. Use of resistant variety, timely sowing/transplanting, indigenous technical knowledge available with the farmers (use of crab traps for *gundhi* bug, Use of fresh *Eupatorium* branches for control of sucking pests etc) optimizes the resource use. Balanced application of fertilizers, appropriate intercultural operations and crop rotations also reduces insect pest and disease problems substantially. Field sanitation and proper weed management helps to reduce the insect pest and disease problem in rice.

Insect Pest Management

The major insect pest of rice in the region includes stem borer, *gundhi* bug, leaf folder, root aphid etc. For management of insect pest in rice spray Monocrotophos 36 EC @ 2 ml/litre or Chlorpyriphos 20 EC @ 1 ml/litre of water at 45 DAT for the control of leaf folder, case worm, stem borer and hispa is effective. Dusting with

Carbaryl 10 per cent or Fenvalerate 0.4 per cent @ 20-25 kg/ha controls *gundhi* bug. Under upland conditions, application of carbofuran 3G @ 1kg a.i./ha controls termites and white grubs. Use of healthy and disease free seed also protects the plants from insect pest and diseases. Growing more than one variety (maintain diversity) is a good option to avoid complete crop failure due to pest/disease problem in a particular variety. Botanicals like neem oil 4 ml/litre of water reduces problem of most of the insect pests. Spraying *Verticilium lecanii* @ 1×10^9 spore/ml reduces the problem of white backed plant hopper (WBPH) in rice. Release of *Trichogramma* egg parasitoids @ 50,000/ha reduces stem borer and leaf folder population to a great extent. Entomophthora fungal infection is severe on white leaf hoppers in North East. The pathogen infects leaf hoppers during booting stage of rice and control the pest population to the tune of 60 per cent. Spraying of *Beauveria bassiana* @ 3 g lit^{-1} of water for the control of **rice hispa** is effective.

Figure 14.22: Damage by Rice Hispa.

Disease Management

Blast, sheath blight, bacterial leaf blight, brown spot, sheath rot are important diseases of rice. Seed treatment with Bavistin 50 WP @ 2 g kg^{-1} seed protects the crop from blast, in nursery and also up to 25-30 DAT. For management of brown spot and blast treat the seed with Dithane M-45 @ 5 g kg^{-1} seed and spray the crop with Tricyclazole 75 WP (0.6 g lit^{-1}). Higher dose of N fertilizer such as urea makes the rice plant susceptible to insect pest and disease. Application of right doses of potassium fertilizer such as MOP improves resistance against insect pest and disease in rice (Figures 14.23 and 14.24).

Physiological Maturity

Harvesting is done at the yellow ripening stage to avoid shattering loss in field. Harvesting should be done when the seeds have attained maximum physiological maturity (in 28 and 31 days, respectively for short and medium duration varieties after the 50 per cent of the spikelets in the panicle have flowered).

Figure 14.23: Symptom of Paddy Leaf Blast.

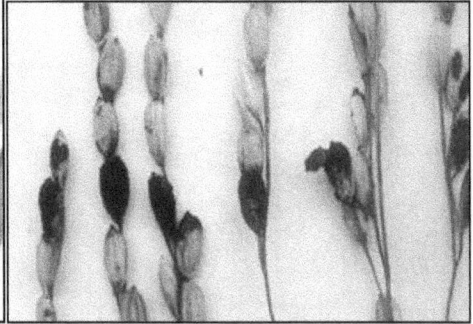

Figure 14.24: Symptom of Paddy Bunt.

Harvesting

When the panicle turns to straw yellow colour the yellowing of plants is activated. When 90 per cent of the panicle are in straw colour with the moisture content of 20 per cent for short and medium duration varieties and 17 per cent moisture for long duration varieties. In hybrid seed production, 'R' line should be harvested first and ensure complete removal from the field followed by harvesting of seed parent. At that stage the irrigation to the seed production plot is withheld and this hastens the drying of the plants/seed. The plants are harvested with intact panicles. The male parent (B/R line) should be harvested first and removed from field and then the seed parent (female) is harvested. Care should be taken to avoid the admixture of female and male lines during harvest (Figures 14.25 and 14.26).

Threshing

The harvested plants are stacked in a cleaned (free from other variety and volunteer plant seeds) threshing floor. Then either by hand beating or with the use of LCT threshers under large scale production for separation from the plants. The preferable moisture for threshing is 15-18 per cent. This will avoid the occurrence of mechanical injury to the seeds.

Drying

The seed should be dried to a safe moisture content of 10-13 per cent under normal drying conditions.

Figure 14.25: Harvesting Male Row (First Removal).

Figure 14.26: Harvesting Female Row.

Grading of Seeds

The chaff, ill filled, under sized and oversized seeds are to be removed to maintain the physical purity of the seed above 98 per cent. It is done through processing. The sieve sizes recommended for different varieties of paddy based on the shape of the caryopsis is as follows.

Size of Seed	Sieve Size
Long slender (Ponni, White ponni)	= 1/16 x 3/4 " (1.3mm x 19 mm)
Slender - IR 50	= 1/15 x 3/4"(1.3mm x 19 mm)
Medium slender (IR 20, CO 43)	= 1/14 x 3/4"(1.5 mm x 19 mm)
Short bold (ADT 36, 37,38,39, TKM 9,Ponmani)	= 1/13 x 3/4" (1.8 mm x 19 mm)

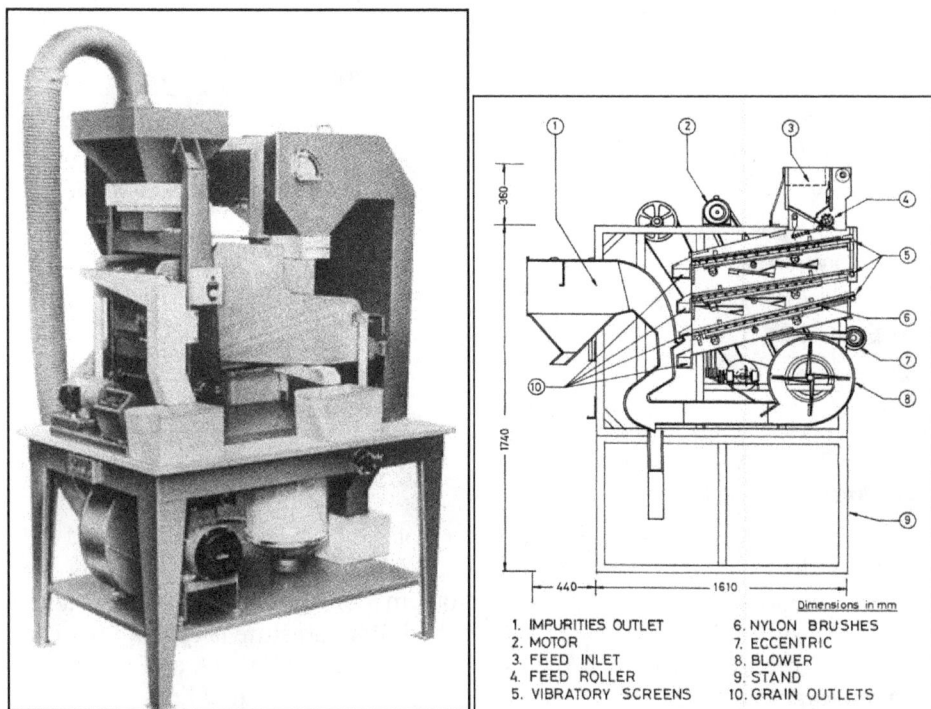

1. IMPURITIES OUTLET
2. MOTOR
3. FEED INLET
4. FEED ROLLER
5. VIBRATORY SCREENS
6. NYLON BRUSHES
7. ECCENTRIC
8. BLOWER
9. STAND
10. GRAIN OUTLETS

Figure 14.27: Air-Screen Cleaner-cum-grader for Size Grading.

The size graded seeds may be upgraded by density grading using gravity separator. Heavy and medium fractions with 90 – 92 per cent recovery could be selected for seed purpose in hybrid seed production,for getting better seed quality, the seeds should be size graded using 1.3 mm x 19 mm oblong sieve and hand cleaning of parental line seeds.

Seed Treatment

The seeds are to be treated with Thiram @ 4 g kg^{-1} or Bavistin @ 2 g kg^{-1} as slurry treatment. For bulk storage, the seeds will be fumigated with celphos @ 3 g m^{-1} in airtight condition for 7 days (or) Decis + Thiram @ 0.04 + 2.5 g kg^{-1} as slurry treatment or with carbendazim @ 2g kg^{-1} of seed using 5 ml of water kg^{-1} of seed (or) dry dress with halogen mixture ($CaOCl_2$ + $CaCO_3$ mixture at 1: 1 ratio) of seed (or) coating the seed with polymer (3 ml kg^{-1} in 5 ml of water) + Royal flow 40 sc @ 2.4 ml kg^{-1} of seeds + imidachloprid @ 6 ml kg^{-1} of seed or exposure of the seeds thrice with 50 per cent CO_2 (4 days for 50 kg container) at 12 per cent moisture content at 15 days interval.

Seed Standards (Hybrids)

Factor	Standards for each Class	
	Foundation	Certified
Pure seed (Min.)	98.0 per cent	98.0 per cent
Inert matter (Max.)	2.0 per cent	2.0 per cent
Huskless seeds (Max.)	2.0 per cent	2.0 per cent
Other crop seeds (Max.)	10/kg	20/kg
Other Distinguishable Varieties (Max.)	10/kg	20/kg
Total weed seeds (Max.)	10/kg	20/kg
Objectionable weed seeds (Max.)	2/kg	5/kg
Germination (Min)	80 per cent	80 per cent
Moisture (Max.)		
Pervious container	13 per cent	13 per cent
Vapour proof container	8 per cent	8 per cent

Storage

For short term storage use gunny bag or cloth bag. The seeds can be stored upto 1-2 year under ambient storage condition without much reduction in germination (80 per cent) provided they are free from rice moth. In moisture vapour proof containers they can be stored for long time provided the initial moisture is below 8 per cent. When compared with varieties, the hybrids and parental lines (A and B lines) are poor in storability. The order of the storage potential of parental lines and hybrid will be restorer line > hybrid > maintainer line > female line.

Mid Storage Correction

To improve the viability and vigour when the germination of seed reduce to 5-10 per cent lesser than MSCS level (Minimum Seed Certification Standard) adopt hydration – dehydration treatment with disodium hydrogen phosphate (3.59 g dissolved in 100 liters of water) to enhance the quality of the seed by 10 to 15 per cent.

Chapter 15

Value Addition and Byproducts in Rice

G. Gurumeenakshi

Post Harvest Technology Centre
AEC and RI, TNAU, Coimbatore, Tamil Nadu
e-mail: phtc@tnau.ac.in

Introduction

Rice is the staple food for 65 per cent of the population in India. It is the largest consumed calorie source among the food grains. With a per capita availability of 73.8 kg it meets 31 per cent of the total calorie requirement of the population. India is the second largest producer of rice in the world next to China. In India paddy occupies the first place both in area and production. Indian Basmati Rice has been a favorite among international rice buyers. Following liberalization of international trade after World Trade Agreement, Indian rice will become highly competitive and has been identified as one of the major commodities for export. This provides us with ample opportunity for development of rice based value-added products for earning more foreign exchange. Apart from rice milling, processing of rice bran for oil extraction is also an important agro processing activity for value addition, income and employment generation. It has been reported that about 9 per cent of paddy is lost due to use of old and outdated methods of drying and milling, improper and unscientific methods of storage, transport and handling. It has been estimated that total post harvest losses of paddy at producers' level was about 2.71 per cent of total production. To minimize post harvest losses, precautions should be taken to follow proper post harvest practises. They include timely harvest at optimum moisture percentage (20 per cent to 22 percent), use of proper method of harvesting; avoid excessive drying, fast drying and rewetting of grains. Ensure drying of wet grain after harvest, preferably within 24 hours to avoid heat accumulation, uniform drying to avoid hot and wet spots and mechanical damage due to handling. The losses in threshing and winnowing can be avoided using better mechanical methods. Proper sanitation during drying, milling and after milling to avoid contamination of grains and protect from insects, rodents and birds and use of proper technique of processing *i.e.* cleaning, parboiling and milling helps in reduction in post harvest losses. To avoid storage losses maintaining optimum moisture content *i.e.* 12 per cent for longer period and 14 per cent for shorter storage period is essential.

Rice Products

☆ **Rough rice or paddy**: Defined as rice in the husk after threshing.

☆ **Stalk paddy:** Defined as unthreshed in the husk, harvested with part of the stalk.

☆ **Husked rice/brown rice**: Rice from which the husk only has been removed retaining still the bran layers and most of the germs. Such rice is sometimes reflected to as bran rice even though there are variations having red or white bran coats.

☆ **Milled rice**: Rice from which husk, germs, bran layers have been substantially removed by lower machinery, also known as polished rice and if milled to high degree it is called as white rice.

☆ **Under milled rice**: Rice from which the husk germs and bran layers have been partially removed by power machinery and is also known as unpolished rice.

☆ **Hand produced rice**: Rice from which the husk, germ and bran layers have been partially removed, without the use of power machinery, also known as "home produced" or "hand milled rice".

☆ **Parboiled rice**: Rice, which has been specially processed by steaming or soaking in water, heating usually by steam and drying. Parboiled paddy can be milled to various degrees or home produced in the same way as ordinary paddy. It is called as parboiled milled or parboiled hand pounded.

☆ **Raw milled**: The paddy, which is milled not after giving heat treatment, such as parboiling.

☆ **Coated rice**: Defined as rice milled to a high degree and then coated with glucose.

☆ **Whole grain**: Refers to husked, milled or hand produced rice which does not contain any broken grains smaller than 3/4 of the size of the whole kernel.

☆ **Broken rice**: Husked, milled or hand produced rice consisting of broken grains of less than 3/4th size of the whole grain but not less than 1/4th.

☆ **Fragmented rice**: Small brokens upto 1/4th size of the whole grain.

☆ **Rice polishing**: Now defined as the by-product from milling rice, consisting of the inner bran layer of the kernel with part of the germ and a small percentage of the stoney interior also known as rice meal or rice flour elsewhere.

☆ **Glutinous rice**: A type of rice, which after cooking has a peculiar stickyness regardless of how it is cooked.

☆ **Scented rice**: A type of rice, which contains aroma and gives scented smell on cooking.

☆ **Rice flour**: Ground polished rice, which is mainly starch with very little gluten. Used in the same way as corn flour and for noodles, sweets and short pastry.

☆ **Quick cooking rice**: Rice that had been partially/completely cooked then dehydrated kernels.

Parched Rice

It is prepared by throwing rice in sand heated to a high temperature in an iron or mud pan. On stirring, rice begins to crackle and swell. Then the content of the pan are removed and sieved to separate the parched rice from sand. Parboiled rice is used for making grayish to brilliant white colour parched rice and sold either salted or unsalted. It is eaten as such or mixed with butter milk or milk.

Puffed Rice (Using Rice)

This popular ready-to-eat snack product is obtained by puffing milled parboiled rice. In the traditional process rice is gently heated on the furnace without sand to reduce the moisture content slightly. It is then mixed with salt solution and again roasted on furnace in small batches with sand on a strong fire for a few seconds to produce the expanded rice. Rice expands about 8 times retaining the grain shape and is highly porous and crisp.

Parched Paddy or Puffed Rice (Using Paddy)

Sun dried paddy is filled in mud jars and is moistened with hot water. After 2-3 min. the water is decanted and the jars are kept in an inverted position for 8-10 hours. Next the paddy is exposed to the sun for a short time and then parched in hot sand as in the preparation of parched rice. Puffed rice is prepared by throwing pretreated paddy into sand heated to a high temperature in an iron pan. During parching the grain swell and burst into a soft white product. The parched grains are sieved to remove sand and winnowed to separate the husk.

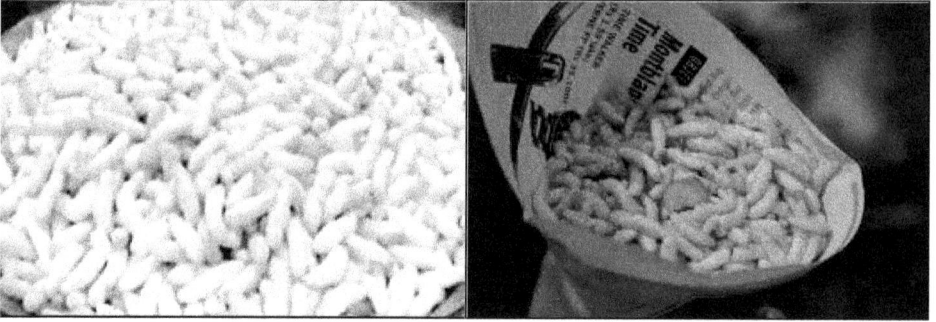

Rice Flakes

The production of rice flakes begins with parboiling of rice which helps to soften the grain and prepare it for processing. Once the rice is tender, the cooked grains are rolled and flattened. After the mixture is the desired thickness, the flattened rice is allowed to dry completely. The dried flakes are run through another rolling process to create simple flakes.

Rice Bars

Snack bars using rice include granola, breakfast, and energy bars. Snack bars target the health conscious consumer and are often enriched. A breakfast bar contains milk infused into the grains composing the bar to give one the taste of a bowl of cereal. Crisped rice is used in granola, breakfast, and energy bars. High-pressure extrusion processing is generally used for the manufacture of crisped rice for these snack bars

Rice Desert

Desserts with rice emerged in the late 1990s with waxy rice serving as a fat replacer in ice cream. The addition of 1.5 per cent waxy rice starch improves the creamy mouthfeel and overall texture, allowing the product to mimic premium (higher fat) ice cream. Crisped rice provides texture to a number of chocolate dessert products.

Idli

Idli is a small, white acid – leavened, and steamed cake made by bacterial fermentation (12-18 hours) of a thick batter made from rice and dehulled blackgram dhal. Idlies are soft, moist and spongy had a desirable sour flavour. For idli, the rice was coarsely ground and the black gram was finely ground. The soft spongy texture observed in the leavened steamed idli made out of black gram is due to the presence of two components, namely surface active protein (globulin) and an arabinogalactan (polysaccharide) in black gram. The mucilaginous principle of blackgram is identified as arabinogalactan. It is believed that this mucilaginous principle helps in the retention of carbondioxide during the fermentation of the thick batter and is thus responsible for the soft spongy honey comb texture of the idli. Fermentation brings about physical and chemical changes in the idli batter. With the progress of fermentation there is an increase in batter volume, acidity and non protein nitrogen.

Dosa

Dosa is another common fermented product used in India. This is prepared from a fermented batter of rice and pulse in the proportions ranging from 6: 1 to 10: 1. Both the ingredients are finely ground, unlike in the idli batter which contains the rice component in a coarse consistency. The dosa batter is very thin and dosa is baked on a hot pan. The thickness of a thin pancake depends upon the consistency of the batter. Thin batter gives a thin pancake, although it may stick to the pan.

Dhokla

Dhokla is a fermented food prepared from rice and bengal gram. This is popular in West India, particularly Gujarat. This is prepared from a batter of coarsely ground rice and bengal gram. The fermented batter is steamed in a pie dish, cut into diamond shape and seasoned.

Extruded Products

Rice based extruded products include sevai, idiappam, murukku (chakli) rice based vadagam etc. Rice based noodles and noodles from fermented rice flour are popular in China, Japan and the orient. Rice flour along with flour of other desirable ingredients (other grain flour, salt, sugar, spices etc.) is mixed and

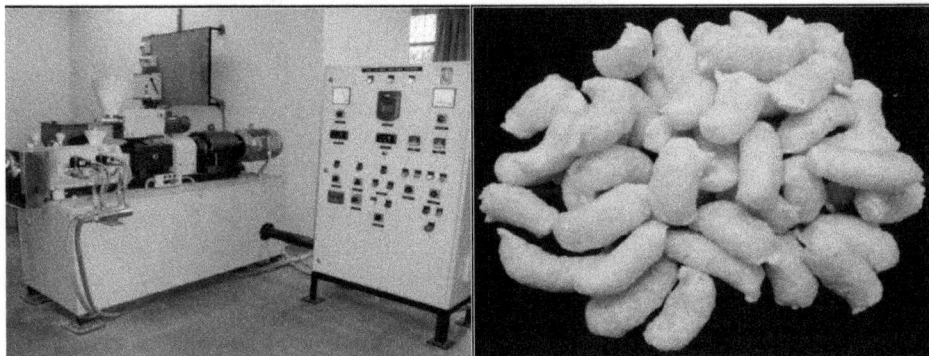

conditioned. Mixed conditioned flour is fed to the hopper of the extruder. After setting temperature of extruder barrel and screw speed, controlled feeding of flour to the extruder barrel is done.For desirable shape of the product, proper die is set in the exit end of the barrel. During the travel of conditioned flour from feed end to exit end, flour is converted into dough mass, cooked and finally when it exits from barrel, there is expansion of product resulting into dried puffed product.

Vadagam

Vadagam is a traditional product mostly prepared from rice flour and also sago. It is a deep fat fried product and consumed as a side dish in the meals. Raw rice flour(100 g), green chillies (1.0 g), cumin seeds(4.0 g) and salt (2.0 g) were taken. The ingredients were mixed and 250 ml of water was added and cooked for 10 minutes until a thick extrudable paste was formed. The hot paste was extruded in hand extruder (dia 5 mm) and dried in sun for about five hours. The dried vadagam sample was packed in a polyethylene bags and sealed.

Preparation of Rice Bread

Breads are made from rice flour, yeast, sugar, egg, fat and water. The mix is made into dough by adding water and kept for rising. Dough is divided into pieces of lime size balls and are allowed to rest for 10-30 minutes. Now it is baked in a proper mould in an oven at 200°C for about 30 minutes. The promising feature of this product is the utilisation of riceflour for making bread. Procedure followed for the preparation of rice bread is same as that of ordinary maida bread. Excellent product could be obtained by replacing 50 per cent maida with rice flour withminor variations in the ingredients. Other baked products like cake, bun and rusk could be developed with rice flour

Germinated brown rice (GBR) is considered whole food because only the outermost layer *i.e.* the hull of the rice kernel is removed which causes least damage to its nutritional value. Brown rice can be soaked in water at 30 °C for specified hours for germination to get GBR. Soaking for 3 h and sprouting for 21 h has been found to be optimum for getting the highest gamma-aminobutyric acid (GABA)

content in GBR, which is the main reason behind the popularity of GBR. The intake of GBR instead of white rice ameliorates the hyperglycemia, boosts the immune system, lowers blood pressure, inhibits development of cancer cells and assists the treatment of anxiety disorders. Germination process could be used as enzymatic modification of starch that affects pasting properties of GBR flour. GBR would improve the bread quality when substituted for wheat flour. It is concluded that GBR has potential to become innovative rice by preserving all nutrients in the rice grain for human consumption in order to create the highest value from rice.

Bran: A by-product from the milling of rice consisting of the outer layer of the kernels with part of germ.

Rice bran oil: It is the oil extracted from the germ and bran of rice.

Rice Bran Products

As rice bran is highly nutritive, easily available and low in cost. If it is properly processed and stabilized, different home made products can be developed useful in improvement of healthconditions. It is concluded that rice bran can be used as a source of fiber and minerals particularly iron

Development of Rice Bran Based Products

Besan ladoo is a common Indian sweet made from gram flour (besan) and sugar. Rice bran based ladoo is prepared by roasting 50g besan and 10 rice bran in 25g ghee and 25g sugar was added finally. Once done forming even bolls were formed.

Chapter 16

Machinery for Paddy Cultivation

B. Shridar and R. Kavitha

Agricultural Machinery Research Centre,
Agricultural Engineering College and Research Institute,
Tamil Nadu Agricultural University, Coimbatore, Tamil Nadu
e-mail: amrc@tnau.ac.in; tnaushridar@gmail.com

Rice is cultivated in 44 million ha in India and is the largest area in the world. During the past 55 years there has been a remarkable increase in the production of rice. The area under rice increased by 1.5 times while the production increased by over four times to the tune of 99 million tones. The productivity has increased from 0. 7 t/ha to 2.40 t/ha. To mitigate the growing population rate rice production should rise to 120 million tones by 2020. This could be achieved only through selective mechanization and increase of productivity as the area is plateaued (Rice Knowledge Management Portal, DRR). The scarcity of manual labour and the drudgery require mechanization to carry out different farming operations.

I. TILLAGE IMPLEMENTS

i. Tractor Drawn Cage Wheel

This is an iron wheel, lugged with L angles. The tractor will not work satisfactorily in puddling of wet lands with the rubber lugged tyre wheel, due to the wheel slippage. To overcome this iron wheels have been introduced. These wheels are of two types *viz.*, half cage wheel and full cage wheel. The iron lugs provide required friction for the movement of the tractor or power tiller and churn the saturated soil simultaneously as the cage wheel advances. The half cage wheels are attached along with the regular rubber tyre wheel. The width of the tractor cage wheel is 1 m and that of the half cage wheel is 0.5 m. The diameter of the tractor power tiller cage wheel depends on the make of the unit. Normally 12 L angle lugs are provided and spaced equally, on each half width of the cage wheel in a staggered manner. Cost of the cage wheels Rs.30,000/- (approx.)

ii. Power Tiller Operated Cage Wheel

The cage wheels are attachments in place of pneumatic wheels to the power tillers and are suitable for pudding operation in rice cultivation in wetlands, where tractor or bullocks find it difficult to work. The cage wheels provide a good traction

and floatation while the rotavator of the power tiller does the puddling. The cage wheel with iron lugs provide required traction for the movement of the power tiller and churn the saturated soil simultaneously as it advances. The cage wheels work well in all the soils except the clay plus silt content of the soil was more than 60 per cent. In addition to saving in cost and time, more uniformity and thoroughness in the puddle is obtained with the cage wheels than by country plough. The average depth of puddle obtained is 23 cm. The capacity of the puddler is 0.44 ha day⁻¹. Normally 12 L angle lugs are provided and spaced equally, on each half with the cage wheel in a staggered manner. Cost of the cage wheels Rs.12,000/- (approx.)

II. SOWING IMPLEMENTS

iii. Tractor Mounted Direct Rice Seeder

It is a tractor drawn implement used for line sowing of rice. Most of the farmers who owns a tractor has a tractor drawn cultivator. Seed boxes along with cup feed type seed metering mechanism are mounted on the cultivator frame and the seeds are dropped in furrows opened by the cultivator shovels. Detachable side wings are fixed to the existing shovel type furrow openers of the cultivator, which helps in placing the seed at the required depth. Power to operate the seed metering discs is taken from the ground wheel drive though a clutch. A square bar is provided at the back of the unit to close the furrows. It saves 70-75 per cent labour and operating time and 60-65 per cent on cost of operation compared to manual broadcasting. It is also reported that there is 16 per cent increase in yield. The width and depth of operation is 1830 mm and 25-35 mm, respectively. The field capacity of the seeder is 0.68 ha h⁻¹. Cost of the unit is Rs.65,000/- (approx.)

Specification

1. Overall dimensions : 2500 x 1030 x 1240 mm
2. Weight : 410 kg
3. No. of rows : 9
4. Row spacing : Adjustable
5. Hill spacing : Adjustable

iv. Improved Direct Rice Seeder

Since rice transplanters require special skills in preparation of mat type nursery, direct rice seeder will be of immense use in rice growing areas especially when water is not limiting factor and labour is the limiting factor. Transplanted rice experiences nursery uprooting shock and total duration is elongated to about 7–10 days. In direct sown rice crop, the transplanting shock is avoided and as a result the crop can be

harvested in lesser duration. The direct rice seeder is useful to sow pregerminated rice seeds directly in wetland without transplanting.

Many Research Institutes have developed direct seeder implement made of MS, PVC etc. Among them, TNAU model of drum seeder has been very popular in certain belts of India.

The rice seeder consists of two/four seed drum hoppers made of PVC, a handle all fixed in a framework made of 16 mm diameter conduit pipe supported by two ground wheels with lugs. The seeder is capable of sowing 8 rows at 200 mm row spacing or 4 rows at 200 mm row spacing. To use this unit the field is to be puddled and leveled at least two days before operating the seeder. The required seed rate is 25 kg ha^{-1}. There is 25-30 per cent yield increase when compared with transplanted rice. The field capacity of the seeder is 1 ha day^{-1}. Weeding and interculture made easy using cono weeders and rotary weeder. Cost of the unit is Rs.4,800/- (approx.)

Specifications

1.	Length (mm)	2000
2.	Width (mm)	1500
3.	Height (mm)	640
4.	Number of rows sown	4 and 8

5. Row to row spacing (mm)   200
6. Number of seed metering holes   9
7. Diameter of the metering holes (mm)   10
8. Number of floats   2
9. Weight (kg)   10
10. Capacity (ha/day)   1.1

v. IRRI Six Row Manual Rice Transplanter

The IRRI-manual rice transplanter is operated using mat type nursery. This is a manually operated pull type implement. Six seedlings mats are kept on the seedling tray of this machine and the operator pushes the handle to enable the picker to pick the pinch of seedlings with soil in the tray in its downward movement and plant the seedlings in the puddled soil. As the handle is lifted back, the picker arm is also taken back from the soil thus completing one stroke of operation. The distance between two rows is 20 cm. The operator walks backward as he pulls forward the transplanter for the next stroke. The distance between two hills within a row is decided by the distance through which the machine is pulled forward in each stroke of operation. The implement is light in weight and can be carried by a single person. Maintenance of this implement is also easy. An area of 0.25 to 0.4 ha can be covered in a day. This implement is ideally suitable for small and marginal farmers. The width of operation is 1000 mm. The depth of planting is 20-25 mm. Cost of the unit is Rs.10,000/- (approx.)

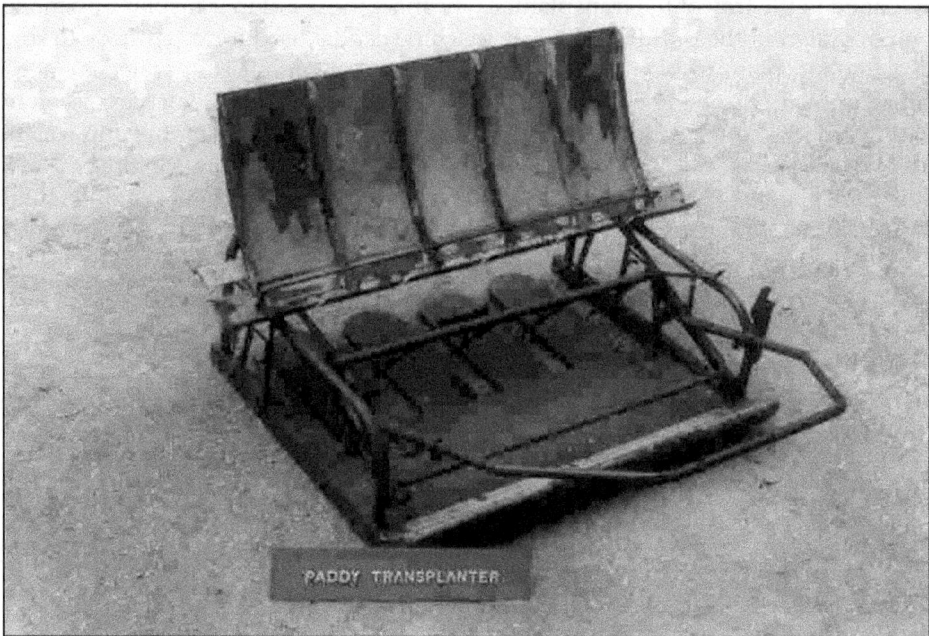

PADDY TRANSPLANTER

Specifications

1.	Length (mm)	1460
2.	Width (mm)	1330
3.	Height (mm)	530
4.	Nursery used	Mat type
5.	Number of rows	6
6.	Row to row spacing (mm)	200
7.	Width of coverage (mm)	1200
8.	Type of fingers	Fixed opening type
9.	Number of persons required	2
10.	Weight (kg)	20 (without seedling mats)

vi. Korean Model Four Row Rice Transplanter (Walk Behind)

It is four row rice transplanter consisting of a prime mover, transmission, engine, float, lugged wheels, seedling tray, seedling tray shifter, pickup fork and pickup fork cleaner. It is a walk behind type rice transplanter using mat type nursery and it transplants the seedling uniformly without damaging them. The planting depth and hill-to-hill spacing can be adjusted. Automatic depth control helps in maintaining uniform planting depth. The machine has safety clutch mechanism, which prevents break down of planting device from the impact against stones in the field. For operation, the machine is transported to the field and mat type nursery is loaded in the tray of the transplanter. The machine is put in transplanting mode and operated in the puddle field. The machine can be used for 30 cm row to row spacing and plant to plant spacing between 14 to 30 cm. It is operated by a 2.3 hp air cooled petrol engine. It is easier to maneuver at turning in comparison to riding type 8 row self propelled rice transplanter. Time loss during turning was 10 seconds in the case of 4 row transplanter whereas it was 20 seconds in the case of 8 row transplanter. The field capacity of the machine is 0.155 ha h^{-1}. Cost of the unit is Rs.2,50,000/- (approx.)

vii. Self-Propelled Rice Transplanter (Yanji Sakthi-Chinese Model)

It is a single wheel driven and fitted with diesel engine. The machine is riding type and it transplants seedlings from mat type nursery in eight rows in a single pass. The drive wheel receives power from the engine through V -belt, cone clutch and gearbox. A propeller shaft from the gear box provides power to the transplanting mechanism mounted over the float. The float facilitates the transplanter to slide over the puddle surface. The tray containing mat type nursery for rows is moved sideways by a scroll shaft mechanism, which converts rotary motion received from the engine through belt-pulley, gear and universal joint shaft into linear motion of a rod connected to the seedling tray having provision to reverse the direction of movement of tray after it reaches the extreme position at one end. Fixed fork with knock out lever type planting fingers (cranking type) are moved by a four bar linkage to give the designed locus to the tip of the planting finger. The field coverage varies from 1.5 to 2 ha/day depending upon the field condition. The machine can be used for 23 cm row to row spacing and plant to plant spacing between 14 to 23 cm. Cost of the unit is Rs.2,00,000/- (approx.)

III. Weeding Machinery

ix. Cono Weeder

The cono weeder is useful for uprooting and burying weeds in between standing rows of rice crop in wetlands. It disturbs the topsoil and increases the aeration. The unit consists of a long handle made of mild steel tube. Two truncated rollers one behind other are fitted at the bottom of the long handle. The conical rollers have serrated projections on the periphery. A float provided in the front portion prevents the unit

from sinking into the puddled soil. The cono weeder can also be used for trampling the green manure crop in addition to weeding operation. They are more efficient than manual pulling of weeds. The cono weeder can satisfactorily remove weeds in a single forward pass with a push pull movement. It is easy to operate by a single operator. The weeder does not sink in puddled soil. The field capacity is 0.18 ha/day. Cost of the unit is Rs.1,500/- (approx.)

x. Single Row Finger Type Rotary Weeder (Japanese Weeder)

The weeder is commercially available with an operating width of 100 mm and is being adopted by the farmers for SRI technique of rice cultivation. The rotary unit has six 'L' shaped blade, the larger leg being 60 mm long and shorter leg being 35 mm long. The shorter leg has four fingers with a throat depth of 20 mm. The rotor is mounted on a frame of 250 mm x 130 mm made of 12 mm flat and is provided with a suitable beam and handle. A float is provided in the front, which can be adjusted according to the height of the operator. The weight of the unit is 3 kg. The field capacity of the unit is 0.2 ha day^{-1}. Cost of the unit is Rs.600/- (approx.)

xi. Motorized Power Weeder

The motorized power weeder comprises engine, gearbox, main frame, rotary wheel, float, handle and controls. The engine and all other accessories were mounted on main frame fabricated out of mild steel pipe. Engine axle is fitted in the final drive mechanism with frictionless roller bearings with packed seal. Throttle is provided on left side handle to control the speed of engine. Two cutter wheels of diameter 300 mm and width 150 mm serve as weeding part. Each cutter wheels had four sharp blades fitted on the periphery. These cutter wheels are fitted on axle and secured. The weeder is suitable for rice at row spacing of 20 to 30 cm. The weeding efficiency of the weeder when used for three row operations in one direction was 62 per cent while the same for two way operation was 77 per cent. When the weeder is used in all rows (two row operation), then the weeding efficiency for one way weeding was 67 per cent and the same for two way operation

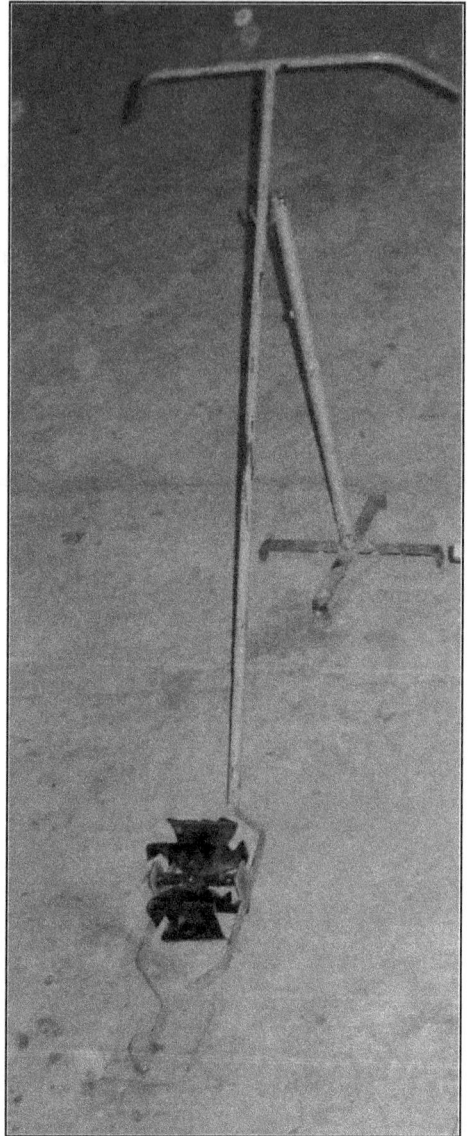

was 88 per cent. Since it is a light weight weeder (17 kg), it can be operated in all soils. The field capacity of the weeder is 0.75 to 1.0 ha per day. Cost of the unit is Rs.40,000/- (approx.)

IV. Harvesting Machinery

xii. Self Propelled Rice Harvester

The harvester is a front mounted unit and consists of a cutter bar, five numbers of gathering head assemblies with star wheels, two numbers of vertical conveyor belts with G.I. pegs on the periphery, a gear box and a pair of cage wheels. The power is taken from the engine pulley through compound idlers to the harvester main shaft. Initially, an area of 3 m x 2 m at a corner of the field and about 30 cm wide strip around the field is to be cut to accommodate the machine and to facilitate free flow of cut crop respectively. The unit can harvest both line and random planted crops. The unit cuts and windrows the cut crop in a straight line for easy collection. The shattering loss is about 1.0 to 1.5 per cent. The unit cannot be used for harvesting crop in lodged condition. It covers an area of 2 to 2.4 ha of land per day of 8 hours. It results in 75.5 per cent of saving in cost and 64 per cent of saving in time as compared to manual harvesting. It is suitable for harvesting all varieties of rice and also can be used for both in line planted and random planted crop. Cost of the unit is Rs.80,000/- (approx.)

xiii. Combine Harvester

It is a machine, which performs the functions of a reaper, thresher and winnower.

Functions

1. Cutting the standing crops
2. Feeding the cut crops with the threshing unit
3. Threshing the crops
4. Cleaning the grains and freeing it from straw
5. Collecting the grains in a container

The functional components are header, reel, cutter bar, elevator, feeder, concave, feeding drum, threshing drum, feeder concave, fan, chaffer sieve, grain sieve, return conveyor, tailing auger, grain elevator and grain container.

Header is used to cut and gather the grain and deliver it to the threshing cylinder. The large combines use T type header with auger tables. However cutting unit does the harvesting which uses the cutter bar similar to that of the mower. The knife has got serrated edge to prevent the straw from slipping while in operation. There is a suitable cutting platform, which is provided with a reel and canvas. The reel is made of wooden slates, which helps in feeding the crops to the cutting platform. The reel gets the power through suitable gears and shafts. The reel revolves in front of the cutter bar while working in the field. The reel pushes the standing crops towards the cutting unit. The reels are adjustable up and down as well as in or out. The cutter bar of the combine operates like the cutter bar of a mower. It cuts the standing crops and pushes them towards the conveyor. The conveyor feeds the crop to the cylinder and concave unit. The threshing takes place between the cylinder and concave unit of the combine. The basic components of the threshing unit of the combine are similar to a power thresher. As soon as the crops are threshed the threshed material move to straw racks. These racks keep on oscillating and separating the grains. The cleaning unit consists of a number of sieves and a fan. The unthreshed grains pass through the tailing auger and go for rethreshing. The clean grain pass through the grain elevator and finally go to the packing unit. Grains are collected in a hopper provided at suitable place. The fan is adjusted such that the chaff etc., blown off the rear side of the machine.

xiv. Mini Rice Combine

The mini rice combine was developed at TNAU, Coimbatore based on commercially available riding type small combine. In this combine, the winnower is modified for effective cleaning of the threshed grain. The riding type small combine consists of two reciprocating cutter bars, one for cutting of ear heads and the other for cutting the remaining stubble. It is driven by a 10 hp diesel engine. The header unit has a four bar reel for guiding the ear heads towards the cutter bar. The converging auger guides the cut ear heads towards the threshing cylinder. A spike tooth cylinder of 325 mm diameter and 385 mm length and a concave and cylinder casing with inclined louvres are provided. The threshed material is cleaned with the help of a screen and an aspirator. The bruised straw is thrown out at the end of the threshing cylinder. A sack holder and platform is provided for bagging cleaned grain. A manual hydraulic pump is provided for raising the header unit and a mechanical lever is provided for adjusting the lower cutter bar. This helps

to harvest crops of different height. The field capacity of the machine is 0.11 ha h^{-1}. Cost of the unit is Rs.2,50,000/- (approx.)

xv. Tractor Operated Combine Harvester (Commercial Model)

This machine consists of cutting, threshing, cleaning and grain handling units. The cutting section includes reel, cutter bar, an auger and a feeder conveyor. Threshing section has wire loop threshing cylinder, concave and cylinder beater. The cleaning section mainly consists of straw walker, chaffer sieve, grain collection pan and blower. The grain handling section consists of grain elevator and a discharge auger. The crop after being cut is delivered to the cylinder and concave assembly through feeder conveyor where it is threshed and grains and straw is separated in different sections. It saves 80-90 per cent labour requirement and 33 per cent cost of operation as compared to traditional method. Cost of the unit is Rs.14,00,000/- (approx.)

xvi. Rice Combine Harvester (Kukje Model)

This combine harvester is available in three and four row models. It is highly suitable for harvesting non-lodged rice under wet soil conditions since they are provided with rubber tracks for locomotion. It consists of crop lifting mechanism with retractable fingers, reciprocating cutter bar, crop conveying system, ear head thresher, winnower and bagging unit. The crop is held between two chain conveyors, threshed and the straw is released in the field in the form of windrows. The cutter bar fixed in front of the unit actually cut the entire crop at ground level and the cut crop is lifted to threshing unit. Before reaching the threshing unit the straw portion leaving the ear head is cut and windrowed in the field. Only the ear head portion is allowed to pass in to the threshing component. In other commercial combines the entire crop is sent to the threshing unit and the straw is cut in to pieces, which leads to difficult to the farmers in collecting the cut straw and using them as cattle field. So in Kukje combines the straw in received as whole without any damage.

The reaping width of the combine is 140 cm and height of cut from ground level is hydraulically adjustable. The grains received from the threshing unit can either be collected in grain tanks or in bags. The maximum capacity of grain storage tank is 1400 litres. The unladen weight of the unit is only 2935 kg which is about 1000 kg less than other combines. The field capacity of the combine is 2 to 3 ha/day. It is powered by a 45-55 hp diesel engine. Cost of the unit is Rs.18,00,000/- (approx.).

xvii. Rice Thresher

A belt attached to an engine or electric motor drives it. The crop is fed to the threshing cylinder through the feeding trough. Rice is threshed due to impact and rubbing action between the threshing cylinder and concave screen. The separated grains falls through the holes of the concave screen. The falling grains are cleaned

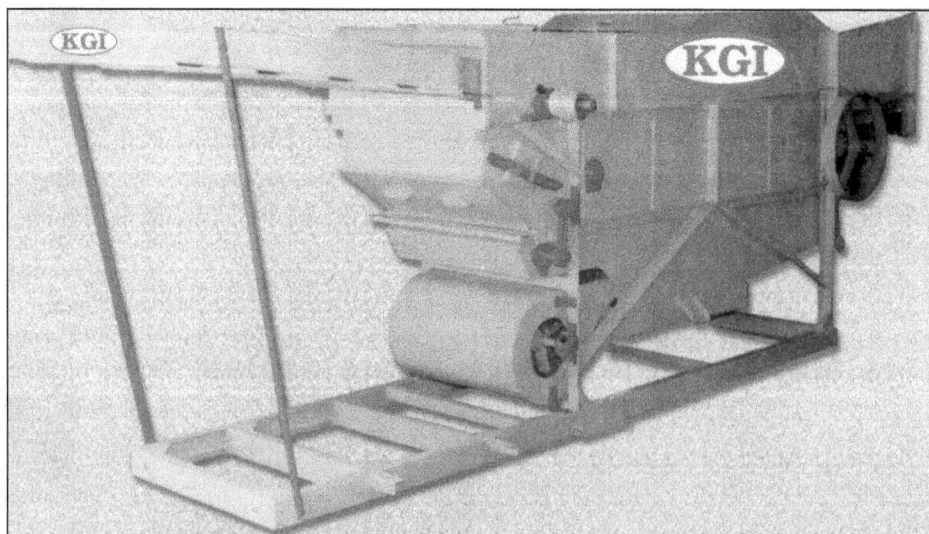

with the help of a fan, driven by a 'V' shaped belt. The clean grain after winnowing goes down through the grain outlet at the bottom of the thresher.

The basic components of a rice thresher are threshing drum, power transmission unit, grain outlet, concave, screen, air outlet, control plate and blower. Threshing drum is a hollow drum having a cylindrical shape made of sheet metal having wire loops on the periphery. The wire loops are usually made of spring wires which are about 4 to 6mm in diameter. The loops are staggered in different rows so that the pairs of leaves of the loops in one row are different from the other rows. The loops are provided with threaded legs so that they can be attached to the drum with nuts on both sides of the drum. A concave screen is provided just below the cylinder.

An oil engine or an electric motor is used to run the thresher by a suitable pulley mounted at one end of the drum. The axle end is connected with the prime mover with the help of a belt. Blower axle receives power from the drum axle. The threshed grains move one side, by means of an auger. The grains are transformed by means of an elevator through a discharge passage when the grains are collected in bags. Concave screen is provided at the bottom support for the crops fed into the threshing chamber. The screen is made of sheet metal with suitable holes, curved in shape of a circle. It extends from the feed through to the rear position of the thresher. A blower is provided extending throughout the width of the thresher sieve. It rotates on an axis parallel to the threshing drum. Number of paddle blades are provided to blow air over the grain. The outlet of the blower is directed towards the thresher so that air blow across the falling grains to separate the chaff. Cost of the unit is Rs.1,50,000/- (approx.).

xviii. Rice Straw Chopper-cum-Spreader

Flail type and cutter bar type rice chopper cum spreader are available. The machine chop the straw left in the field after harvesting in to pieces and spread it in a single operation. The chopped stubbles will be buried in the soil by two operations of disc harrow or one operation of rotavator and decayed after irrigation. The machine consists of a rotary shaft mounted with 38 flails fitted in three rows to harvest the straw. The chopping unit consists of a cylinder with 6 rows of 23 serrated knives each and four counter rows with 22 knives. The width of cut is 2.28 m. The straw after cutting by the flail, pass on to the chopping mechanism. The field capacity of the machine is 0.44 to 0.52 ha h^{-1}. Cost of the unit is Rs.3,50,000/- (approx.).

Chapter 17

System of Rice Intensification

B.J. Pandian

Water Technology Centre,
Tamil Nadu Agricultural University,
Coimbatore, Tamil Nadu
e-mail: directorwtc@tnau.ac.in

SRI was developed by Father Henry de Laulanie who was striving to improve the livelihood of the poor rice farmers of Madagascar. Prevailing situations forced the farmers to use younger seedlings and less water and Father Henry de Laulanie observed that these cultivation environment lead to more growth of rice. Triggered by this, he started to develop a new method of rice cultivation and ended up with System of Rice Intensification (SRI) which resulted in extraordinary yield gains.

It took 20 years before SRI was made known to the rest of the world, mainly due to the persistent initiative of Dr. Norman Uphoff of Cornell University. Now that it has spread to more than 20 countries and replete with innumerable success stories, efforts are on to generate and establish the exact scientific mechanisms responsible for the observed beyond belief or too good to be true successful SRI results.

Principles of SRI

SRI encourages rice plant to grow healthy with large root volume, Profuse and strong tillers, Non lodging, More and well filled grains per panicles and higher grain weight.

SRI is based on the following principles:

☆ Seed rate of 7.5 kg ha^{-1}

☆ Mat nursery

☆ Young seedling (14-15 days) for planting

☆ Single seedling per hill

☆ Square planting with wider spacing (25 x 25 cm)

☆ Rotary weeding up to 40 DAT at 7-10 days interval starting from 10 DAT

☆ Irrigation after disappearance of ponded water

☆ Nitrogen management through Leaf Colour Chart (LCC)

SRI Practices

SRI is a boon to the farmers in terms of decrease in monetary and non-monetary inputs and increase in yield. SRI involves less expenditure on fertilizers and plant protection chemicals. In SRI, rice crop grows healthy in natural conditions and its root growth is massive. It receives nutrients from deeper layers of the soil. Maximum tillering (30 tillers/plant can be easily achieved; 50 tillers per plant are quite attainable) occurs concurrently with panicle initiation. Under excellent management even 100 fertile tillers per plant or even more can be achieved due to early transplanting and absence of die back of roots.

The principles of SRI are achieved by following certain practices. These are explained hereunder.

Young Seedlings

SRI calls for use of younger seedlings. Based on preliminary trials, 14 day old seedlings were recommended in Tamil Nadu for SRI. At this stage there will be 3 leaves in the plant. If the nursery bed is properly prepared with sufficient organic manure, the seedling growth will be good to handle. Seedlings of 8-12 days are also used by farmers.

Besides this, younger seedlings do not experience transplanting shock. When aged seedlings are planted there will be transplanting shock which extends to a week sometimes without much growth of the plants. So, the avoidance of transplanting shock period also helps to add up to the total number of tillers. Another reason for higher tiller density is the increase in the number of days for tillering. As the plants go to the main field 10-20 days in advance.

SRI Nursery

Since single seedling is planted per hill at wider spacing, the number of seedlings required for planting is drastically reduced (Tables 17.1 and 17.2, Figure 17.1).

Only 5 - 7.5 kg of seed is required to plant 1 hectare is the first benefit accruing to the farmer. Since the nursery area gets reduced from 800 m² to 100 m² and the need to maintain the nursery is only for 14 days, the nursery costs reduces considerably (second benefit).

Table 17.1: Number of Seedlings Required to Plant 1 Hectare

Nature of Variety	Hill Spacing (cm)	Number of Hills per sq.m	Number of Seedlings per Hill	Number of Seedlings per sq.m
		SRI		
All varieties	25 x 25	16	1	16

**Table 17.2: Nursery Area, Seed Requirement and Seedling Density
in the Nursery (to plant 1 hectare)**

Method	Nursery Area (sq.m)	Seed Requirement (kg)
SRI	100	7.5

Figure 17.1

Since the seedlings have to be planted at 14 days or less it is necessary to raise them properly. For this, raised bed nursery method has been proposed.

SRI nursery could be prepared according to the soil conditions. But, it is important to lay the nursery such that seedlings could move to the main field and planted in the shortest possible time. So, it would be ideal to have the nursery in one corner of the main field.

The nursery (to plant 1 hectare) has to be prepared in 100sq.m. It is recommended to prepare 20 beds of 1 x 5 m each. The beds of 5 cm height could be prepared by scooping out the soil around the beds so that furrows are formed all round the beds. Above the bed, a plolythene sheet or used fertilizer bags have to be spread. Over the sheet, again 4 cm soil bed has to be prepared. In case the native soil is not so fertile, 95 grams of DAP for each 5 sq.m may be mixed with the soil before spreading it. Bamboo rods could be laid on the sides of the beds to prevent sliding of the soil.

For each 5 sq.m bed, 375 g of pre-germinated (2 days) seeds (weight prior to soaking) have to sown evenly. The furrows could be irrigated to soak the beds. Alternately, rose cane could be used to spray water. The nursery bed could be prepared in wet field or dry field depending upon the situation. If too much sunlight is expected, nursery could be covered with coconut fronds for two days.

☆ If farm yard manure is applied to the nursery soil, it should be a well decompose done, otherwise, seedlings will be scorched.

☆ If the seedlings do not attain good growth to plant in one week time, urea @ 0.5 per cent may be sprayed.

Pulling Out of Seedlings

It would be ideal to remove the seedlings along with the soil intact with the roots and plant them immediately. This helps the seedlings to establish quickly. In the beginning this might look cumbersome. But handling lesser number of seedlings; nearness to the planting area (if nursery is in the same field); and repeated tries will ward off this feeling. Farmers adopt different ways to remove seedlings according to their convenience. If it is found difficult to get the seedlings along with soil, they may be pulled out from the bed and bundled as done in conventional way but without giving much trauma to the seedlings.

In many places, the same labourers employed for planting are used to remove the seedlings from the bed without involving additional cost.

Main Field Preparation

Good leveling of the main field is essential in SRI. Otherwise, young seedlings planted in depressions will die. Farmers generally use wooden planks etc., to level the puddle field. Laser leveling by tractor is appropriate if available for hiring. Tamil Nadu Department of Agriculture is contemplating to introduce laser levelers for SRI. Farmers who could not achieve proper leveling may tie a brick in a rope and drag along depressions to drain stagnating water.

Field drainage is an important component in SRI. It is also recommended to provide small drains on all the sides of the main field. Drainage channels are formed along with raised beds for SRI in some places.

Nutrient Management

Organic manures are recommended in SRI cultivation, since they are found to give better response. Enrichment of soil for nutrient supply with tank silt ($40 - 50$ t ha^{-1}), FYM/compost (15 t ha^{-1}) and incorporation of *in situ* grown 45-60 day old green manures such as sunhemp/*dhaincha* are ideal for basal incorporation. However, since all rice farmers are not in a position to adopt organic farming, Integrated Nutrient Management (INM) is currently recommended. INM is not merely application of fertilizers along with organic manures. It is the application of available organic sources like, cattle manure, poultry manure, vermicompost, green manures, green leaf manures, biofertilizers and supplementing with fertilizers in adequate splits to meet the nutrient demand of the crop at different growth stages.

Plant Spacing

In SRI, seedlings have to be planted at wider spacing of 25 x 25 cm is recommended.

Table 17.3: Spacing and Plant Density

Method of Cultivation	Plant Spacing (cm)		Plant Density (m^{-2})	
	Short Duration Crop	Medium/Long Duration Crop	Short Duration Crop	Medium/Long Duration Crop
SRI	25 x 25	16		

Figure 17.2

The tillering rate and number of tillers produced per plant will depend upon the space between plants. The wider spacing concept in SRI produce more tillers. The reasons for higher tiller density in SRI are : (1) since younger seedlings are used they have higher vigour to produce tillers and additional number of days to produce tillers, (2) competition between plants for light and nutrients is reduced due to wider spacing, (3) soil churning by weeder has a specific positive effect on the growth of the plants.

Number of Seedlings Per Hill

In SRI, only one seedling has to be planted per hill. If the main field has been leveled properly, all the seedlings will establish well. Otherwise seedlings in depressions where water stagnate may die.

Square Planting

Square planting at 25 cm spacing is recommended for SRI could be achieved by several means. A commonly practiced method is to use either coconut fibre or nylon rope to guide the line of planting. Small sticks or colour clothes are inserted at 25 cm spacing to show the place of plating within the row. Markings are done with paint also. The rope has to be held by two persons who can move it after every

25 cm row is planted. If a line of seedlings are planted at the two sides of the field the guide rope could be easily operated. Markings may also be made in a bamboo stick or aluminium pipe.

Figure 17.3

A roller marker made of steel rods developed by a farmer in Andhra Pradesh has become a very useful tool in SRI now. When the roller is pulled over the level main field, cross lines are formed on the soil and the intersecting points indicate the place for planting the seedlings.

Figure 17.4

Single seedling has to be planted at the intersecting point. In recent analysis in the TN-IAMWARM project showed that farmers engaged 45-75 labourers for conventional planting and 22-48 labourers for SRI planting, the average being 60 and 35 respectively.

That the seedlings should not be planted deeper than 2 cm in the soil is an already existing recommendation. But, in reality the seedlings are planted much deeper resulting in poor tillering as the tiller producing buds get buried.

The main purpose square planting is to use the weeder in both directions. The primary benefit in SRI is obtained by the weeder use only. For conventionally planting labourers, square planting might look cumbersome. But they will soon realize they have to plant only 25 per cent of seedlings they would normally handle.

Water Management

In SRI irrigation up to panicle initiation stage, it is recommended to irrigate the field to 2.5 cm after the previously irrigated water disappears and hairline cracks develop. After panicle initiation, irrigating to 2.5 cm depth one day after the previously ponded water disappears from the surface. At hairline crack stage soil will not be dry but it will still be moist.

The shallow irrigation can save water up to 40 per cent (Table 17.4) and there will not be any yield loss due to this.

Table 17.4: Research Finding on SRI Water Management (Coimbatore, 2001)

	Recommended Practice	SRI
Irrigated water (m^3 ha^{-1})	16634	8419
Per cent water saved	–	49.4

A genuine apprehension of the farmers would be that the limited irrigation might lead to weed infestation. Since weeder operation commences in 10-12 days and done every 10 days thereafter, weeds do not have the opportunity to come up. The important point farmers should remember is that rice does not require flood water and it is enough to keep the soil moist. Farmers using ground water will realize the water, time and electricity saved by SRI irrigation. If SRI is adopted in an entire command area, water saved will be sufficient for use in other areas or purposes.

Weeder Use

Using weeder is of primary importance in SRI. The tool is called weeder, because it was introduced three decades ago only for weed control. But the weeder use in SRI has multivarious effects and benefits.

The difference in weed control adopted in conventional cultivation and SRI is given in Table 17.5.

Figure 17.5

Table 17.5: SRI Weed Control Methods

Method	Labour Required (ha⁻¹)
Using weeder in both directions at 10-12, 20-22, 30-32 and 40-42 DAT	30

When SRI was introduced, two types of weeders were recommended : cono weeder and rotary weeder. Cono weeder weighs about 7.5 kg and can be operated only by men labourers. Rotary weeder is lighter (about 2 kg) and could be operated by women labourers also.

☆ The first weeder use at 10-12 days is crucial in SRI and should not be missed

☆ Herbicides are not recommended in SRI

☆ Some water should be there in the field while using the weeder

☆ It is important to remove the left out weeds by hand

☆ Some earthing up takes place when the weeder used. This makes the plants to produce new roots which add up the root activity

☆ The suitability of the type of weeder is site specific and depends upon soil conditions and labour mindset

☆ It would be ideal to manufacture the weeder by an ironsmith in the nearby area by getting a suitable model.

Varieties Used for SRI

It has been an universal observation that any variety, whether high yielding or land race, shows a higher response under SRI. The genetic potential of any variety or cultivar is expressed better in SRI because of the changed growing environment. Cultivars, like ASD16 in Tamil Nadu, which were considered as shy tillering, produced 100 per cent more of tillers under SRI. However, hybrids seem to fair well when compared to other cultivars.

Figure 17.6

Effect of SRI Practices

The changes brought about in the growing environment of the plants due to the use of young and single seedling, wider spacing, weeder operation and limited irrigation have been found to have positive influence on the growth of plants as well as in the soil dynamics in a manner not found in conventional cultivation.

Crop Growth

- ☆ The seedlings establish quickly without transplanting shock
- ☆ Profuse root growth and roots remain white. In flood irrigated rice, roots become brown and less active after panicle initiation
- ☆ Profuse tillering
- ☆ Leaves remain green till harvest accruing higher photosynthetic activity during grain filling
- ☆ SRI crops do not lodge
- ☆ Higher number of panicles, number of grains per panicle and less sterility.

Pest Scenario

☆ Differences exist in varietal response to insect damage under SRI.

☆ Reduction in major pests like stem borer, whorl maggot and leaf folder was observed with SRI when compared to normal wetland rice

☆ Rat damage nil or less.

Grain Yield

It is estimated that if the SRI adopted even on just 25 per cent the irrigated rice area in India (5.3 mh), there would be a saving of paddy seed worth Rs. 500 crore. The increase in production because of increased yield (40 per cent) and better water productivity (32 per cent water saving) bringing more area under irrigated rice cultivation is estimated to be about five million tonnes- enough food grains for about four million families for a year under the public distribution system.

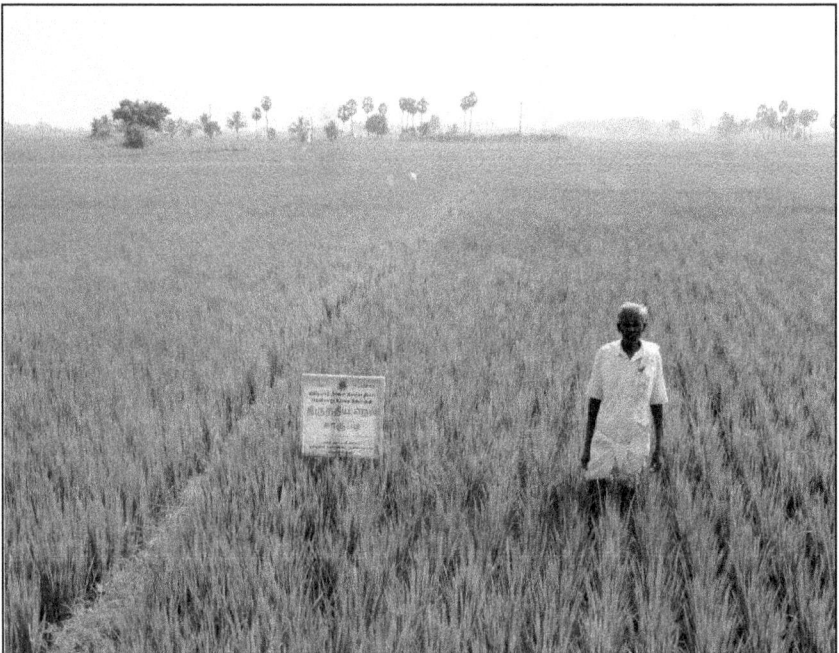

Figure 17.7

The expected increase in net income would be about Rs. 6,741 crore. Assuming that 50 per cent of the irrigation is groundwater-based, SRI would save energy (40 per cent) equal to 632.61 million kWh which means a savings of Rs. 253 crore, as well as 10,000 million cu.m of irreplaceable groundwater for future generations. This is apart from unaccounted benefits such as health, nutrition and general improvement in the rural economy.

Chapter 18

Organic Rice Cultivation in Tamil Nadu

S. Ramasamy and R. Kuttimani

Department of Sustainable Organic Agriculture,
Tamil Nadu Agricultural University,
Coimbatore – 641 003, Tamil Nadu
e-mail: organic@tnau.ac.in

General Guidelines for Organic Farming

Organic farming is a crop production method respecting the rules of the nature, targeted to produce nutritive, healthy and pollution free food. It maximizes the use of on - farm resources and minimizes the use of off – farm inputs. It is a farming system that seeks to avoid the use of chemical fertilizers and pesticides. Commitment to protect and preserve nature is a pre-requisite for practicing organic farming. In organic farming, the entire ecosystem (*i.e.* plant, animal, soil, water and microorganisms) is to be protected.

The general guidelines on organic production of crops are prepared based on National Programme for Organic Production (NPOP) launched by Government of India. These guidelines enable the growers to attain more or less the same level of the productivity of conventional farming within a few years and at the same time maintain the fertility of the soil and protect the ecological balance.

It is essential that all the crops in the organic field follow an organic method of production. Though the farmers are free to convert a portion of the farm, it is advisable that the entire farm is converted to organic. However, in the case of large farms, the conversion can be phased out for which a conversion plan is drawn out and followed systematically. The time from the start of organic management to the certification of crops and/or animal husbandry is known as the conversion period. The whole farm including the livestock should be converted according to the

standard over a period of three years. A simultaneous production of conventional, in conversion and/or organic crops or animal products, which cannot be clearly distinguished from each other, is not allowed. To ensure a clear separation between organic and conventional production, a buffer zone or a natural barrier should be maintained. An isolation distance of at least.

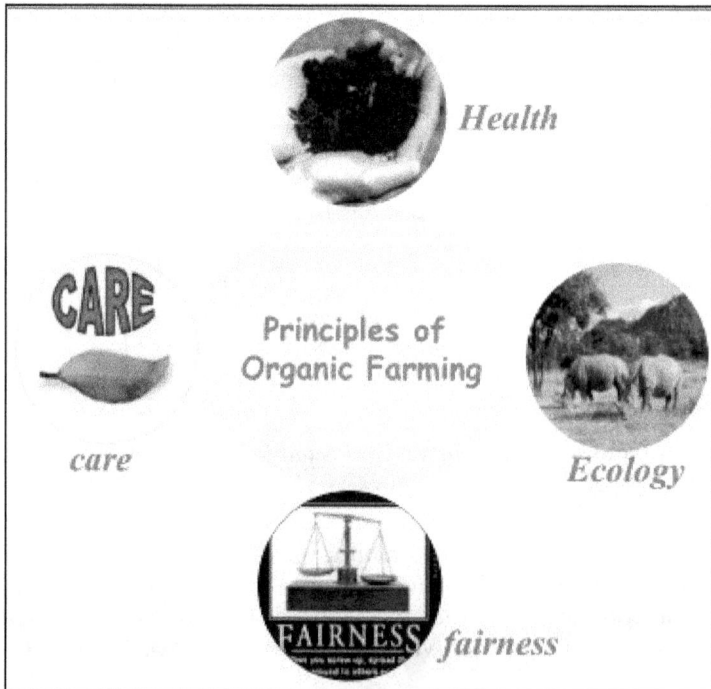

A buffer zone of 25m width is to be left from all around the organic. The produce from this isolation belt shall not be treated as organic. Mixed farming systems integrating crop husbandry and livestock are the most ideal where the livestock is also maintained following organic standards. This would enable the use of by-products in the farm itself without depending on external sources.

Organic Rice Cultivation

Rice can be cultivated under a variety of climatic and soil conditions. Rice cultivation is conditioned by temperature parameters at the different phases of growth. The critical mean temperature for flowering and fertilization ranges from 16 to 20°C, whereas, during ripening, the range is from 18 to 32°C. Temperature beyond 35°C affects grain filling. More uniform and warm conditions enable more than one crop to be taken per year in this region of Tamil Nadu.

Rice-based Organic Farming System

A model rice based organic farm is illustrated below. Crop components are integrated with of the farm wastes and allied activities suitable for the farm and the region. In general recycling and biodiversity are the key components in an organic farm.

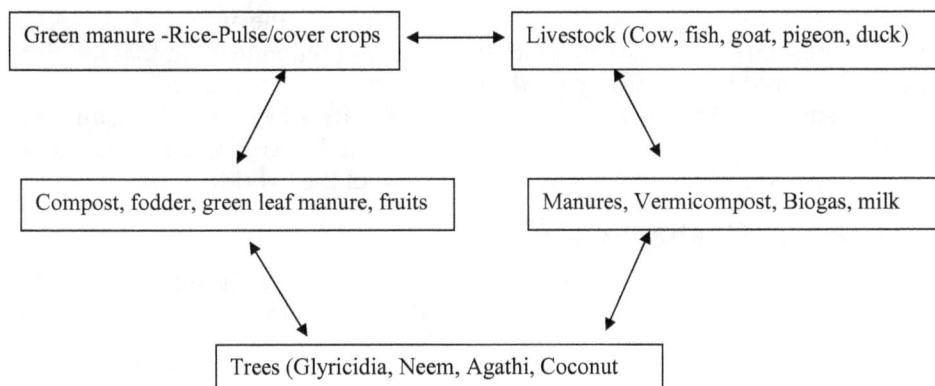

Cropping System

Crop rotation is an agronomic practice followed by farmers to make use of nutrients present in the soil in the best possible way. When the same crop is planted every season the soil becomes deficient in a particular nutrient that is utilized largely by that plant. This situation can be prevented by cultivating crops that have different nutrient requirements. When leguminous plants are cultivated, they trap the atmospheric nitrogen and convert it into a form that can be easily utilized by the plants. When the root nodules and leaves of these plants get into the soil, they increase its nitrogen content and help to retain the soil fertility. Cultivating *Sesbania* as an intermediate crop between two rice crops gives good results. The following sequence can be a viable option for sustainable organic rice production in the tropical climate of Tamil Nadu.

Jun-Aug	Aug-Jan	Jan-Mar	Apr-May
Green manure	Rice (medium duration)	Pulse	Cover/poly crops

The crops are planned in such a manner that the nutrients used by the first crop should be replaced by the following crop. The nutrient requirement of the second crop should be different. This will help greatly to maintain the nutrient balance in the soil. The cover crop raised during hot weather period can prevent the soil organic carbon loss and maintain soil microbes intact for better nutrient mobilization.

Soil

Top soil should be ideally 18–23 cm deep. While cultivating rice, it is always good to study the type, nature and the nutrient content of the soil before adding nutrients. This can be done by having soil samples tested in a soil-testing laboratory. Manures can be applied based on the nutrient status of the soil. In rice cultivation, the yield will be high when the pH of the soil is between 5 and 6.5. The yield will be poor if the pH of the soil is below 5 or above 9. Alluvial soil, sandy clay and clayey soils are suitable for rice cultivation.

Variety

The variety chosen should not only be suitable for the agro-climatic region but also resistant/tolerant to the pests and diseases predominant in that location. The seeds should be organically produced. Low nitrogen requiring varieties with good consumer preference is the basis for the selection of variety for organic rice. Hybrids require more nutrients, its selection need to be scrutinized based on the farm resource recycling capacity to meet the need of the hybrid.

Selection of Quality Seeds

Seed selection plays an important role in rice cultivation. The seeds selected for cultivation should be of uniform size, age and free of contaminants. They should also have good germination capacity.

Seed Rate

The seed rate varies according to the space, method of stand establishment and season from as low as 5 kg to 60 kg.

Seed Treatment

Seed treatment helps to improve germination potential, vigour, and resistance to pests and disease. The different methods of rice seed treatment are:

1. **Seed treatment with *Pseudomonas fluorescens*:** Treat the seeds with talc based formulation of *Pseudomonas fluorescens* 10g/kg of seed and soak in one litre of water overnight. Decant the excess water and allow the seeds to sprout for 24hrs and then sow the seeds in the nursery.

2. **Seed treatment with biofertilizers 1:** Five packets (1kg/ha) each of *Azospirillum* and Phosphobacteria or five packets (1kg/ha) of Azophos

bioinoculants are mixed with sufficient water wherein the seeds are soaked overnight before sowing in the nursery bed (The bacterial suspension after decanting may be poured over the nursery area itself).

3. **Seed treatment with biofertilizers 2:** Biofertilizers like Azospirillum/ azotobacter/pseudomonas (@ 1.25 kg/ha) are first mixed in one litre of cooled rice gruel. Spread the sprouted seeds on a clean floor, add the biofertiliser slurry and mix well. The mixing of seed and biofertilizers slurry can be done in a pot as well. Dry the seeds in the shade for 30 minutes before sowing. Drying the seeds for half an hour in the bright sun before sowing improves germination and seedling vigour.

Nursery Management

Nursery may be wet or dry or modified mat (*dapog*). Area required varies according to the method of nursery. A maximum of 800 m^2 is required to plant one hectare of main field.

Optimum Age of Seedlings for Quick Establishment

☆ Optimum age of the seedlings is 3 to 4 weeks after sowing.

Pulling Out the Seedlings

☆ Pull out the seedlings at the appropriate time (4th leaf stage).

☆ Pulling at 3rd leaf stage is also possible. These seedlings can produce more tillers, provided enough care taken during the establishment phase through thin film of water management and perfect levelling of main field.

☆ Transplanting after 5th and higher order leaf numbers will affect the performance of the crop and grain yield. Then they are called as 'aged seedlings'. Special package is needed to minimize the grain yield loss while planting those aged seedlings.

Root Dipping

Prepare slurry with 5 packets (1 kg/ha) each of *Azospirillum* and *Phosphobacteria* or 5 packets of (1 kg/ha) *Azophos* inoculants in 40 litres of water and dip the root portion of the seedlings for 15 – 30 minutes in bacterial suspension and transplant.

Main Field Management

Keeping the field suitable for rice production is the major task of the system. The crops suggested in the cropping system mentioned above need to be oriented systematically for successful organic rice cultivation with minimum external inputs. Once the crop rotation as given above is followed without much deviation, then the field preparation for rice cultivation becomes handy. Once the soil becomes

friable then very minimum disturbance as 'tillage operation' is needed. The crops cultivated as cover crops during hot weather period can be incorporated and then green manure is raised. Once the green manures attains flowering, the field is given flooding followed by puddling with cage wheel. Otherwise green manure-trampler drawn by animals is the solution to incorporate the biomass of green manure.

Transplanting

The rice seedlings are transplanted @ 2 seedlings per hill at a depth of 2-3 cm. Shallow planting ensures quick establishment and more tillers. Deeper planting (> 5cm) leads to delayed establishment and reduced tillers. The spacing between the seedlings will vary according to the variety cultivated. Line planting permits rotary weeding and its associated benefits. Allow a minimum row spacing of 25 cm to use rotary weeder. The seedlings may be planted at 10 cm between plants. Fill up the gaps between 7th and 10th DAT.

Weed Management

Weeding can be done by either manual or mechanical means. The weeds are buried alive while using rotary weeder. If removed manually the weeds should be trampled into the field for *in situ* conservation of nutrients and for organic matter as mulch. The first weeding should be done at about 15–20 days after transplantation. Subsequent weeding should be done as and when weeds appear and become problematic.

Managing Soil Fertility

In addition to soil nutrition build or inducted into the soil trough crop rotation and green manures *in situ* cultivation before the rice planting add all the crop nutrients to the soil and enriches the soil physical and chemical activities.

Farmyard Manure

Wastes of cattle, goat and pig are generally used as farmyard manure. All the nutrients required by the plants are present in small quantities in this manure. They remain in the soil for longer periods and produce good results.

> ☆ In addition to green manure incorporation, FYM or compost or sheep/goat manure poultry waste or pig dung can be added to the field but well before (7-10 days before) transplanting.

> ☆ Green leaf manures available in the farm can also be supplemented to the field which are low in soil fertility.

Green Manure Crops

Crotalaria juncea

Sesbania rostrata

Cowpea

Sesbania aculeata

Cluster bean

Biofertilizer Application

Some of the microbes that are commonly used in rice cultivation for the purpose include azotobacter, azospirillum and phosphobacteria. They not only reduce the cultivation cost of using chemical fertilizers but also increase the yield and improve the fertility of the soil.

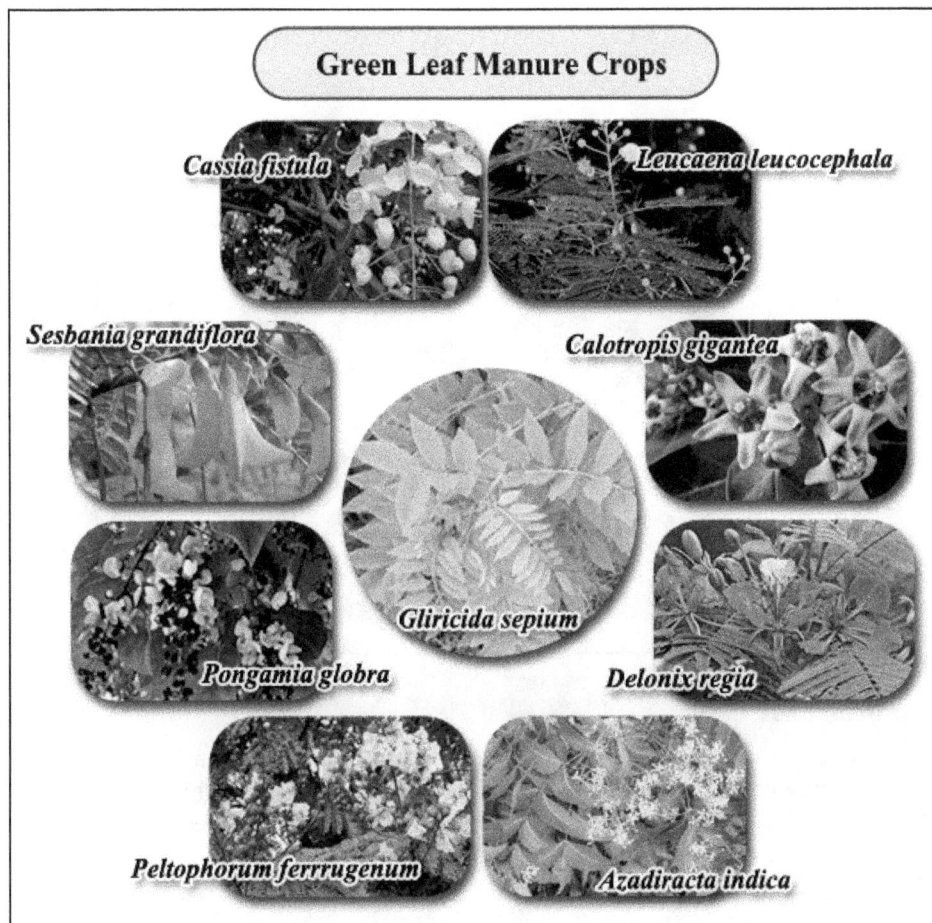

Green Leaf Manure Crops

Cassia fistula — *Leucaena leucocephala* — *Sesbania grandiflora* — *Calotropis gigantea* — *Gliricida sepium* — *Pongamia globra* — *Delonix regia* — *Peltophorum ferrrugenum* — *Azadiracta indica*

☆ Mix 10 packets (2 kg/ha) each of *Azospirillum* and Phosphobacteria or 10 packets (2 kg/ha) of Azophos inoculants with 25 kg FYM and 25 kg of soil and broadcast the mixture uniformly in the main field before transplanting.

☆ *Pseudomonas fluorescens* (Pf 1) at 2.5 kg/ha mixed with 50 kg FYM and 25 kg of soil and broadcast the mixture uniformly before transplanting.

☆ Broadcast 10 kg of soil based powdered BGA flakes at 10 DAT or Azolla as a dual crop by inoculating 250 kg/ha 3 to 5 DAT and then incorporate during weeding or both can add soil fertility if possible and available.

Split Application of Manures to Organic Rice

Rice can also be applied with split dose of organic manures. If the growth of rice is not satisfactory, a booster dose as vermicompost, or oil seed cakes or both may be made if economical.

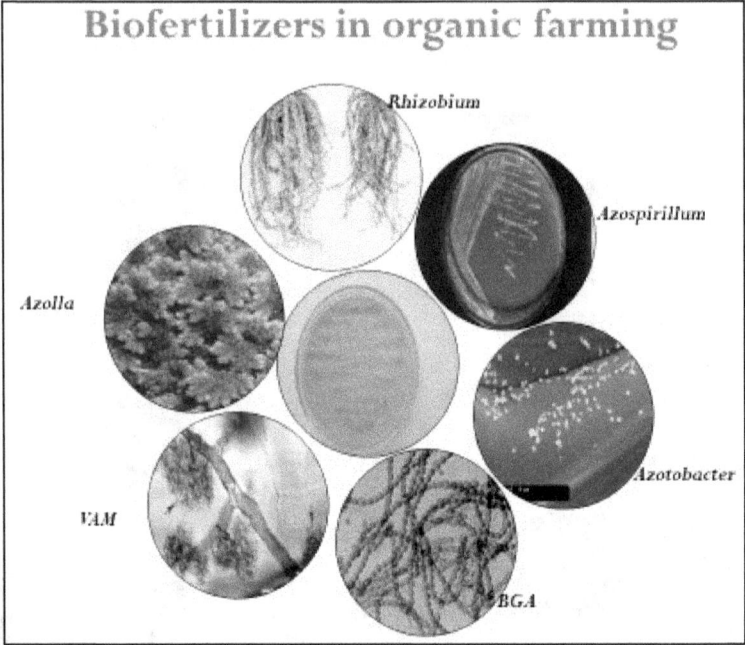

Vermicompost

To decide the time and dose of vermicompost, Leaf Colour Chart (LCC) method may be followed, when the LCC score goes below three. A minimum quantity of 1000kg/ha may be applied at a time. This process may be repeated if necessary. It should be noted that the vermicompost produced from the farm alone is economical.

Oil Seed Cake

There are different kinds of oil seed cakes available such as groundnut cake, neem seed cake and castor seed cake. Generally, neem and groundnut cakes are used for rice.

Finely Powdered Oilseeds Cake Required (kg per hectare)

Cake	Basal Manure	Top Dressing
Neem seed cake	150	60
Groundnut cake	100	25

Plant Growth Regulators

Panchagavya

This is a growth regulator produced from a combination of five products obtained from the cow fermented along with a few other bio-products. For coarse varieties, two sprays of 3 per cent panchagavya during tillering and booting stage should be given. For fine varieties, one spray of 3 per cent at booting stage is sufficient.

Panchakavya: wonder pesticide/biofertilizer

- Soak the groundnut cake in water, before mixing all ingredients listed.
- Keep container covered to avoid dust and dirt
- Stir the contents of container daily
- The panchakavya gets ready in one week
- Spray @ 300 ml of concentrate to 10 litres of water.

Water Requirements

Thin film water management during stand establishment and restoration of irrigation a day after disappearance of ponded water given to a depth of 2.5 cm between tillering and panicle initiation and 5 cm standing water during reproductive phase is best for rice growth.

Pest Management

Leaf Folder (*Cnaphalocrocis medinalis*)

☆ Spray 5 per cent neem kernel extract.

☆ Release ducks in the field to feed on the pests.

☆ Install bird perches like 4–5 branches or twigs of fishtail palm or wild Saccharum in the field to attract predatory birds.

Rice Case Worm (*Nymphula depunctalis*)

☆ Larvae found in the field can be controlled using the rope method.

☆ Drain water from the field for 3-4 days or apply raw cow dung to the standing water. This prevents the respiration of the larvae which is normally through rectal gills.

☆ Burn old worn out bicycle tyres during evening hours (light traps).

Caterpillars: Green horned caterpillar (*Melanitisledaismene*), Yellow hairy caterpillar (*Psalis pennatula*), Army worm (*Spodoptera litura*), Skipper (*Pelopidas mathias*).

☆ Neem cake is applied as basal manure can minimize the problem.

☆ A simple way to keep away caterpillars is to have neem leaf bunches placed at different places within the field. These should be placed in at least twenty five places per hectare.

☆ The caterpillars can be easily controlled by using wood ash. For every hectare, 25 kg of wood ash should be mixed with sand and strewn in the field. This is a low cost technology which can be followed even by a small farmer.

☆ The field should be flooded and then drained. By this method, the larvae and pupae concealed in the soil can be exposed and removed.

Gall Midge (*Orseolia oryzae*)

☆ Spread fresh leaves of *Cleistanthus collinus* (@ 10 kg leaves/100 m² area) in the field at the initial stage of infestation.

Short-horned Grasshopper (*Hieroglyphus banian*)

☆ Spread Calotropis leaves beside the bunds of the field to prevent the entry of grasshoppers.

☆ *Sesbania aegyptiaca* can be grown as a hedge around the field.

☆ Brush the affected crop with branches of Boswellia serrate and place its twigs in the field at a distance of 6–8 m. This should be done in the evening after irrigation.

☆ A solution of cow dung or goat dung can be used. Take about 30–50 kg of the dung and put this in a gunny bag. The gunny bag is balanced on a pole. Below the gunny bag a drum is kept filled with 100–200 litres of water. The tip of the gunny bag should be kept in such a way that it touches the surface of the water. The gunny bag is shaken twice a day for 15 days. After 15 days the water in the drum will be brown and a foul smell will emerge. This should be diluted with twice the amount of water and sprayed. It acts as a repellant for grasshoppers.

Green Leaf Hopper (*Nephotettix virescens*)

☆ Before transplanting, treat the seedlings in neem seed kernel extract for 24 hours. This increases pest resistance.

☆ Neem oil and pongam oil should be mixed in the ratio of 1 : 4 and sprayed. The favourite egg laying spots of the pests – like wild grasses and weeds – should be removed from the field and bunds.

Brown Plant Hopper (*Nilaparvata lugens*)

☆ Use of high levels of nitrogenous fertilizers favours the increase of the BPH population. Hence such fertilizers should be used judiciously.

☆ Water stagnation should be avoided.

☆ Light traps can be used to monitor and attract BPH adults. The trapped insects can be killed

☆ Field and bunds should be cleaned of weeds thoroughly as these function as an alternate host for the insects.

☆ BPH can be controlled by practising the traditional 'neekal podum murai'.

☆ The crops should be planted with proper spacing.

☆ Leaf extract of Lasiosiphon eriocephalus (nachinaar)is effective in controlling BPH. One kg of the leaves is boiled in ten litres of water, filtered, diluted with water in the ratio of 1 : 10 and sprayed, once during the nursery stage and again after transplantation. Leaves of *Calotropis gigantea* can be spread in the interspaces and worked into the field.

Mealy Bug (*Heterococcus rehi*)

☆ Neem seed kernel extract can be sprayed. Burn the straw of Paspalum scrobiculatum and Echinochloa frumentacea near the affected field. Insects get drawn and die. They can also be picked physically and killed.

Earhead Bug (*Leptocorisa acuta*)

☆ The Cycas (*Cycas cercinalis*) flower – called sannampu in Tamil – is used to deal with this bug. It is cut and tied to a stick that is taller than the crop level. The stick is then placed along with straw in 10–15 places in the field. This arrangement repels the adult ear-head bugs and prevents their entry into the field for about two weeks. By this time, the milky stage is over and the crop attains maturity.

☆ *Achyranthes aspera* root (25 kg) and neem bark (12.5 kg) should be dried and powdered. This should be mixed with sufficient quantity of water and sprayed. This quantity is sufficient for one hectare of the crop.

☆ Broadcast kerosene (one litre) mixed with rice bran (4–5 kg).

Black Bug (*Scotinophara coarctata*)

☆ Apply neem cake as top dressing to control the entry of beetles.

Yellow Stem Borer (*Scirpophaga incertulas*)

☆ The land should be ploughed immediately after harvest to destroy eggs and pupae.

☆ Apply neem cake (42–50 kg) as basal manure.

☆ Neem cake bags can be placed in the irrigation channel.

☆ Trichogramma cards can be used. The egg cards of the parasitoids, *Trichogramma japanicam, Trichogramma presiliensis, Telenomus pelefecience* are available commercially.

☆ These parasitoids are capable of destroying the egg masses of stem borers.

☆ Male adult moths can be attracted and trapped using pheromone traps. Seven to eight pheromone traps should be used per hectare. By this method the population of the insect pest can be controlled.

☆ The adult moths can also be attracted using light traps and then destroyed.

☆ The seedlings should be planted with proper spacing.

☆ Put branches of *Erythrina indica* in the field (3–4 m apart).

☆ Spray turmeric rhizome extract. Spread leaves and stems of Datura in the field and allow them to decompose. They act as repellents.

Fungal Disease

Rice Blast (*Pyricularia oryzae*)

☆ Crush the bark of *Careya arborea* (2–3 kg) and apply to the field.

☆ Boil one kilogram of tulsi leaves in two litres of water. Strain and spray the solution twice at 15 days interval (@ 2 ml per litre of water). Leaves of wild tulsi plants can be used for this purpose.

☆ Seed Treatment with TNAU Pf 1liquid formulation @ 10 ml/kg of seeds

☆ Seedling root dipping with TNAU Pf 1liquid formulation (500 ml for one hectare seedlings)

☆ Soil application with TNAU Pf 1liquid formulation (500ml/ha)

☆ Foliar spray with TNAU Pf 1liquid formulation @ 5ml/lit

Rice Brown Leaf Spot (*Helminthsporium oryzae*)

☆ Seed treatment with 20 per cent mint leaf extract for 24 hours.

☆ Spread the mature leaves of *Cleistanthus collinus* all over the field (25 quintals/hectare) and allow them to decay. Irrigate after three days.

☆ Dusting of ash checks the spread of this disease.

Sheath Blight (*Rhizoctonia solani*)

☆ Seed Treatment with TNAU Pf 1liquid formulation @ 10 ml/kg of seeds Seedling root dipping with TNAU Pf 1liquid formulation (500 ml for one hectare seedlings)

☆ Soil application with TNAU Pf 1liquid formulation (500ml/ha)

☆ Foliar spray with TNAU Pf 1liquid formulation @ 5ml/lit

Bacterial Leaf Blight (*Xanthomonas campestris* PV. *Oryzae*)

☆ A slurry is prepared by mixing 20 kg of cow dung with 200 litres of water. The mixture is trained through a gunny bag. The filtrate is further diluted with 50 litres of water and allowed to stand. The water is then decanted, strained and sprayed.

Harvest

The crop should be harvested when the grains are fully mature and turn yellow or straw colour (varies according to the variety). Timely harvesting ensures good grain quality and consumer acceptance, since the grain is less likely to break when milled. Harvesting should be carried out when the moisture content of the grain is about 20–25 per cent and when about 80 per cent of the panicles have about 80 per cent of ripened spikelets. If delayed, grain may be lost due to damage by rats, birds, insects, shattering and lodging.

References

Crop Production guide. 2012. Published by Depart. of Agriculture, Govt. of Tamil Nadu, Chennai and Tamil Nadu Agrl. Univ., Coimbatore- 3. pp. 3-50.

Defining organic agriculture. 2014. Evaluating the Potential Contribution of Organic Agriculture to Sustainability. Natural Resources Management and Environment Department, Food and Agriculture Organizations Document Repository (FAO). In. http://www.fao.org.

Organic farming. 2014. TNAU Agritech Portal. In. agritech.tnau.ac.in

Package of Practices for Organic Cultivation of Rice, Groundnut, Tomato and Okra. 2006. Prepared by Centre for Indian Knowledge Systems (CIKS), Kotturpuram, Chennai-600 085, Tamil Nadu. pp. 9-59.

Principles of organic agriculture. 2014. The organic information hub of International Federation of Organic Agriculture Movements (IFOAM). In. http://infohub. ifoam.org/en/what-organic/definition-organic-agriculture.

Ramasamy S. 2014. Rice: Science and Technology, Kalyani Publishers. New Delhi, India. 342p.

Ushakumari, K. and V.L. Gethakumari. 2009. The adhoc package of practices recommendations for organic farming. The Directorate of research, Kerala Agricultural University, Thirssur. pp. 1-25.

Chapter 19

Speciality Rices of Kerala

B. Shanmugasundaram[1], P. Raji[2], Veena Vigneswaran[3] and K. Karthikeyan[4]

[1]Associate Professor (Agricultural Extension); [2]Associate Professor (Plant Pathology)
[3]Assistant Professor (Plant Breeding); [4]Assoicate Professor(Entomology)
Regional Agricultural Research Station(RARS),
Kerala Agricultural University,
Pattambi, Palakkad District, Kerala

Rice is one the three most important food crops in the world and it is the staple food for over 2.7 billion people. In India area under rice is 44.6 m ha with total output of 80 million tonnes(Paddy) with an average productivity of 1855 Kg/ha. It is grown in almost all the States. West Bengal, Uttar Pradesh, Madhya Pradesh, Bihar, Odisha, Andhra Pradesh, Assam, Tamil Nadu, Kerala, Punjab, Maharashtra and Karnataka are major rice growing states and contribute 92 per cent of area and production. Rice is the staple food for the people of Kerala and it is grown in a gross area of 2.08 lakh hectares producing 5.68 lakh tonnes with a productivity of 2733 kg/ha (2011-2012 Farm Guide 2014). Rice in Kerala is grown in varied ecological situations and is a home for vast diversity of cultivated and traditional land races and wild forms. High rainfall on an average of 3228 mm coupled with undulating topography, interlinked rivers and their drastic formations, backwater systems and saline water inundation from the Arabian sea have created a variety of heterogeneous environments where rice alone could survive and farmers from centuries have selected with their wisdom and experience a diversity of cultivars suitable for all conceivable ecosystems (Sahadevan, 1966).

These varied ecosystems include rice cultivation which extends from about 3 meteres below mean sea level (Kuttanad of Aleppy district) to 1,400 meteres above mean sea level (Vattavada area in Munnar). Different situations such as Modan (purely rainfed uplands) Palliyals/Myals(single cropped terraced uplands), double cropped wetlands(both transplanted and semi dry conditions), Kuttanad area (flooded conditions), Kole and Pokkali (Saline soils), deep ill drained regions, Onattukara (sandy area), Poonthal padam (marshy conditions) and high altitude regions (Sasidharan *et al.*, 2003). Varieties ranging from 80 to 200 days are cultivated in diverse agro-climatic in all seasons Viruppu (autumn) (April/May- September/October), Mundakan (Winter) (September/October to December/January) and in Punja (Summer) (December/January to March/April).al land races and wild forms. High rainfall on an average of 3228 mm coupled with undulating topography, interlinked rivers and their drastic formations, backwater systems and saline water inundation from the Arabian sea have created a variety of heterogeneous environments where rice alone could survive and farmers from centuries have selected with their wisdom and experience a diversity of cultivars suitable for all conceivable ecosystems (Sahadevan,1966). These varied ecosystems include rice cultivation which extends from about 3 meteres below mean sea level (Kuttanad of Aleppy district) to 1,400 meteres above mean sea level (Vattavada area in Munnar). Different situations such as Modan (purely rainfed uplands) Palliyals/Myals(single cropped terraced uplands), double cropped wetlands(both transplanted and semi dry conditions), Kuttanad area (flooded conditions), Kole and Pokkali (Saline soils), deep ill drained regions, Onattukara (sandy area), Poonthal padam (marshy conditions) and high altitude regions (Sasidharan *et al.*, 2003). Varieties ranging from 80 to 200 days are cultivated in diverse agro-climatic in all seasons Viruppu (autumn) (April/May- September/October), Mundakan (Winter) (September/October to December/January) and in Punja (Summer) (December/January to March/April). On account of immense diversity in farming system in Kerala the state is blessed with enormous genetic resources.Rice is one of the very few crop species endowed with rich genetic diversity. Those varieties that differ from the typical staple varieties in their quality features are defined as 'speciality' rices, which include glutinous (waxy) rices, aromatic basmati and non-basmati rices including jasmine rices, and colour rices. Besides the widely known Basmati rices of the Indian subcontinent, jasmine(Khao-Dhak-Mali) of Thailand and waxy (Puttu rices of India, Heinue of China, etc.,) rices there are many less known varieties of unique quality definable as speciality rices (Siddiq, 2004).

India as a member of World Trade Organisation(WTO) enacted the Geographical Indications of Goods (Registration and Protection) Act 1999 which has come into force with effect from 15 Sep 2003. A Geographical indication(GI) in relation to goods means an indication which identifies such goods as agricultural goods, natural goods or manufacturing goods, as originating or manufactured in the territory of a country or a region or a locality in that territory, where a given quality, reputation or other characteristics of such goods is essentially attributable to its geographical origin and in case where such goods are manufactured goods, one of the activities of either the production or of processing or preparation of the goods concerned takes place in such territory, region or locality. In history, unique products from different parts of the world became popular in association with specific geographical names and gained consumer acceptance world wide because of the uniqueness and quality that the product has. Champagne, Swiss watches, China Silk, Darjeeling Tea, Basmathi Rice, Kanchipuram Silk, Malabar pepper etc., are examples for such products with world wide fame. In India 215 unique products have been registered as GIs till date and nearly 200 products are awaiting registration(Elsy *et al.*, 2014). Darjeeling tea from West Bengal, Kangra Tea from Himachal Pradesh, Laxmanbhog Mango, Khirsapati Mango and Fazli Mango from West Bengal and Naga Mircha from Nagaland are some of the registered agiricultural GIs from Northern part of India. Through GI registration, the producers can get market protection for their products which enhances market demand, adding to better livelihood security.

These speciality Rices of Kerala which were accorded Geographical Indication(GI) tag due to the efforts taken by Centre for Intellectual Protection, Kerala Agricultural University, Kerala and various famers Association is discussed in this chapter. The speciality rices having GI tag from Kerala include:

1. Navara rice
2. Palakkadan Matta rice
3. Pokkali rice
4. Wayanad Jeerakasala rice
5. Wayanad Gandhakasala rice and
6. Kaipad rice

Navara Rice

Navara is being called by seven different but closely resembled names(Navara;Navira; Njavara; Njavira; Nakara; Namara and Nakara Puncha) in different regions of the state. Since the name, Navara has more close similarity with the Sanskrit name Nivara, Navara is used commonly It is a unique indigenous rice cultivator of the State having medicinal properties.The Rice is known in the State from time immemorial and according to traditional farmers and healers it is a precious gift from God to the "God's own country" –Kerala.(Anil Kumar 2004). Buchnan (1807) has recorded that Njavara was grown in Kerala in the early 19th century. It is grown in Palakkad, Malappuram, Kozhikode, Waynad, Kannur, Thrissur, Ernakulam, Kottayam and Allepy distiricts of Kerala. Navara is of two types, the white glummed (husked) and black glummed, In Kerala, black glumed Njavara types and yellow glumed Njavara types are being cultivated and the latter correspond to white glumed Njavara mentioned in Ayurvedic literature. The white (yellow) variety is popular in the southern district of Kerala, while in Northern districts ; the practitioners use the black variety (Leena Kumary, 2004). The 12th century Ayurvedic text, Ashtangahridayam describes the white Navara as medicinally superior but the black glummed variety seems to be preferred by physicians today. Njavara genotypes are in general, tall, lanky plants with erect and glabrous pale green-to-green leaves, green basal leaf sheath. Navara is a early duration variety with about 60-90 days. Ayurveda states, 'the rice grains, especially of early duration varieties are acrid, sweet, oleaginous tonic, aphrodisiac, fattening, diuretic, improves taste, useful in biliousness, and relieve the thridosha-Vata, Pitha, Kapha". Navara plants generally grow up to 1m tall. Grain yield of Navara ranged from 919 to 2684 kg/ha (Reddy, 2000). The Variety is known as Shashtikam in Sanskrit due to its extra short duration.Navara has an important place in the Ayurvedic system of treatment and is used alone or in combination with other herbs to treat many diseases.Marunnu Kanji or Mukkudi Kanji a rice gruel made of Navara along with few species and medicinal herbs is considered as best dish to prevent various diseases during monsoon season.Navara rice is recommended as a safe food for diabetic patients and weaning food for infants. It is a cure- for haemorrhoides,urinary complaints,stomach ulcer and Polio. It is also used as Aphrodisiac, muscle builder and for snake bites (Anil Kumar, 2004). Navara rice bran oil is used for a wide range of aches and painful conditions like the cervical spondylosis, low back ace, paralysis, rheumatoid arthritis.Navara is used in Navarakizhi a speciality treatment from Keralas traditional medicine used for curing neuro muscular disorders. Oleation to head and body using special cloth pieces containing a smooth paste of Navara rice cooked in Sida (Sidarectusa. Lin) decoction and milk, makes the body supple, removes stiffness of joints due to various vitiated Vatha dominant conditions, cleans the body channels, and bring blood circulation. It improves complexion, increases apetite, improves digestion, restores relish for food, and corrects the mental irregularities. This makes the body strong and steady, rejuvenated with well developed musculature, Judicious application of this is very effective in hypertension, skin diseases and prevents premature ageing.(Leena Kumary 2004. Njavara has also anti –cancer properties. Scientists of the Kerala Agricultural University (KAU) have isolated a gene from Njavara to beat carcinogens. Molecular studies revealed the presence of Bowmem-Brik inhibitor(BBI) protein in Njavar, which is effective especially against breast cancer (The Hindu dt. 15.08.2007).

Palakkadan Matta Rice

Also known as Rosematta rice or Kerala Red Rice or Red Parboiled rice is an indigenous variety of rice grown in Palakkad district of Kerala. It is different from brown rice and is popular in Kerala and Srilanka. Kerala Matta rice has been historically popular due to its rich and unique taste. The rice is mentioned in Tamil classics such as Thirukkural. PalakkadanMatta got the Geographical Indication (GI) tag for being a rice variety with a distinct taste. Under the registry, there are 10 varieties of the PalakkadanMatta *viz.*, Aryan, Aruvakkari, Chitteni, Chenkazhma,Chettadi, Tavalakkanna, Illapappoochampan, Poochamban, Vattan, and new varieties Jyothi and Kunjkunju. It is a coarse variety of rice with bold grains and red pericarp. The rice has a unique taste. The grains are yellowish pink from being parboiled with the reddish outer layers. Rosematta rice maintains a pink hue as well as its flavor on cooking, Like all brown or parboiled rice, Rosematta has a lengthy cooking time and requires extra water. The coarse rice with red pericarp by itself ensures high content of nutrients. "Par-boiling" of the rice further ensures retention of nutritional value. These varieties are mainly cultivated in black cotton soils as well as in "Poonthalpadam soils" (Soil is heavy, containing 60-80 per cent of clay and silt and possess low permeability and high water holding) of Palakkad district.

Pokkali Rice

Sasidharan *et al.*, (2003) reported three types of saline soils are recognized in Kerala, based on their location, extent and intensity of salinity and crop season. These

Palakkadan Matta- Jyothi

Pokkali rice- Vytilla 1

soils account for 26400 ha. The Pokkali soils are the predominant in this category. It is situated in the coastal lines of Ernakulam and Alappuzha district. Pokkali lands are known after the renowned Pokkali rice variety, which is internationally accepted as a gene donor for salt tolerance in rice. The peculiar reclamation method followed to make this saline soils suitable for rice cultivation is known as 'Pokkali cultivation'. Pokkali soils are acidic with a pH range of 3.1 to 4.8. Padmaja *et al.,* (1994) reported that Pokkali soils are deep, dark bluish black in colour, impervious and clayey in texture, which form hard mass with cracks on drying and turn sticky on wetting. During summer months the saline water from the Arabian sea inundates the Pokkali fields and thus they become saline. The rice varieties/cultivars cultivated in Pokkali tract are called Pokkali varieties/cultivators. Cheruvirippu, Chettivirippu, Karuka, Ponkaruka, Karutha Karuka, Mundakan, Anakodan, Eravanpaddy, Orkayama and Orppandy are some of the Pokkali rice cultivars.According to Sasidharan (2004) Pokkali rice varieties are tall with lodging nature.(Pokkali in local language (Malayalam) means the one who stays tall(Pokkam = tall and Ali= that stays) Most of these varieties are poor yielders with yield range of 1 to 1.5 t/ha. However, the five high yield saline tolerant varieties released(Vyttila-1, Vytilla-2, Vytilla-3, Vytilla-4 and Vytilla-5) from Rice Research Station, Vyttila of Kerala Agricultural University with built in tolerance to salinity and to other abiotic stresses like deep water and soil acidity have a yield potential of up to 4.5 t/ha. Rice is cultivated in these soils during rainy season from June when the salinity level of the water in the fields is low. In order to survive in the water logged field, the rice plants grow up to two months. But, as they mature, they bend over and collapse with only the panicles standing upright. Harvesting takes place by end- October only the panicles are cut and the rest of the stalks are left to decay in the water, which in time becomes feed for the prawns that start arriving in Nov-Dec. Then the second phase of the Pokkali farming, the prawn filteration begins. The organically grown Pokkali is famed for its peculiar taste and its high protein content.

Kaipad Rice

The saline soils of the coastal belt in Kannur districts comprising of the marshy lands where salt water intrusion is a constant problem are classified as Kaipad lands. It comprises saline prone coastal wet land rice production tracts in Kozhikode and Kasaragod district. The Kaipad farming system is similar to the Pokkali farming method – in which a single crop of rice is grown and is alternated with aquaculture in salt water wetland. This farming method is carried out in a natural way, relying upon the monsoon and sea tides. Paddy cultivation is followed by traditional fish farming in these tracts, during the high saline phase (November The volume to April). Kaipad rice is red in colour and it is non-sticky and tasty. Traditional varieties such as Orkayma and Kuthiri are popular in Kaipad fields.Ezhome 1, Ezhome 2 and Ezhome 3 are the first high yielding and non-lodging organic red rice varieties, for the saline prone Kaipad rice fields of Kerala, with awn-less, non-shattering grains and favourable cooking qualities better than local cultivars. The average yields of 'Ezhome varieties varies from 1.4 to 2.0 t/ha These new seed varieties have been developed by the Kerala Agricultural University with the participation of farmers of Ezhome panchayat. This is the first venture of Kerala Agricultural University

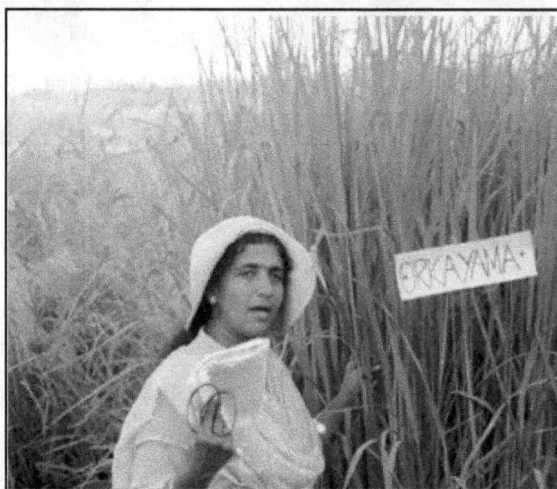

Kaipad Rice

in development of a variety i) adopting the combined strategy of Global Breeding (traditional), and Participatory Plant Breeding, and ii) carrying out all stages of variety development directly in the problem area of farmers

Waynad Gandhakasala and Jeerakasala

The Waynad district of Kerala which has high ranges is famous for the cultivation of traditional scented varieties. Low temperature of 19° c prevailing in this area is suitable for scented rice cultivation. Scented rice cultivation is not popular with farmers because of their low yield and also due to lack of rubberized mill facilities. Gandhakasala and Geerakasala are the most well known indigenous

Jeerakasala Rice

Jeerakasala Rice

Gandhakashala Rice

scented rice of Waynad area. These types grow tall and lodge.They have very long growth duration and are weakly photosensitive but are not very susceptible to pests and diseases. These varieties are patronized by the farmers mainly for their own consumption (Leena Kumary *et al.*, 2003).

Waynad Jeerakasala Rice

This scented non basmathi rice is famous for its characteristics fragrance and aroma. Grains are straw to dull greyish colored, slightly longer and slender grains, aromatic with white kernel. They are weakly photosensitive and suitable for transplanting during virippu(First crop). Wayanad Jeerakasala rice differs from Basmati rice due to growth habit, areas of original cultivation, phsico-chemical properties of grains and grain shape,. The grain are long bold with average grain yield of 2.74 t/ha.Jeerakasala is a tall variety (130-140 cm height) with long duration (150-180 days) lodging with thin culms. Protection of Plant varieties and Farmers right Authority, Government of India honoured the Kuruma and Kurichya tribal Communities of Waynad district through "Plant Genome Savior Community Award" in 2008 for their contribution in the conservation of traditional rice varieties including Jeerakasala.The uniqueness of this rice is mainly attributed to particular climatic conditions prevalent in the area, together with varietal characters and system of rice cultivation, adding to the best expression of aroma and flavor in the product. This aromatic rice is used for the preparation of special food like ghee rice, also called 'Neichore'.

Waynad Gandhakasala Rice

Gandhakasala is another scented rice variety grown in Wayanad. The average yield of Ganhakasala rice is 2 to 2.7t/ha. The variety is of 150-180 days duration with a plant height of 150-155 cm.Grains are straw colored, small, round grains, sandol flavoured with white kernal.They are weakly photosensitive suitable for transplanting during virippu season(First crop). It is mainly used for the preparation of special dishes like ghee rice/fried rice or Neichore prepared for marriages and festivals and also for preparation of sweet items like 'Payasam'. The uniqueness of the product is maintained by adopting Organic method of cultivation. For better Aroma the crop is raised in the Nancha season(Winter season) in Wayanad.

Conclusion

Kerala the God's own country is generally regarded as the Spice Abode of India. But the GI tags given to some speciality rices of the state show that there is a better scope for export orientation of rice from this small state which in turn will improve the socio economic status of the small farmers growing rice. However it is generally seen that farmers awareness in Geographical indication is not in the desired levels even though many consumers of speciality rices are ready to give premium prices to the product. Hence a specific programme on Market extension of speciality rice need to be augmented. With the State Government recent declaration of converting Agriculture to Organic farming by 2016 these speciality rice would get more impetus under the programme. The state governments initiative of giving premium prices for farmers growing speciality rice need to be further strengthened by increasing the area of the programme and providing market support to the farmers through the Supply Co. (Paddy procurement organization of State Civil Supply Corporation).

References

Anil Kumar, N. 2004. Role of Navara (Njavara) *Rice in Traditional healing and health care system in Kerala*. Paper presented In: Science Society Interface on Medicinal and Aromatic Rices organized by Regional Agricultural Research Station, Pattambi on 20-21, August 2004 at Pattambi, Kerala.

Buchanan, F.1807. A journey from Madras through the countries of Mysore, Canara and Malabar. Reprinted in 1988 by Asian Educational Sciences, New Delhi, India 1: 424 p.

Elsy, C.R., Jesy Thomas, K. and Anson, C.J. 2014. *Geographical Indications for enhancing Livelihood of Farming Communities*. In: 7th National Extension Education Congress Organised by Society of Extension Education, Agra on Nov 08-11, 2004 at ICAR Research complex for North Eastern Himalayan Region, Umiam, Meghalaya. pp. 35-41.

Farm Guide. 2014. Farm Information Bureau, Government of Kerala

Leena Kumary, S., Susamma George, P., Dijee Bastian, Aipe, K.C. and Rema Bai, N. 2003. Aromatic rices of Kerala. In (eds.) Singh R.K, and Singh U.S."A treatise on the scented rices of India", Kalyani Press, New Delhi, pp. 317-326.

Leena Kumary,S. 2004.*Genetic Diversity, characerisation and evaluation of medicinal rice in Kerala*. Paper presented In: Science Society Interface on Medicinal and Aromatic Rices.organized by Regional Agricultural Research Station, Pattambi on 20-21, August 2004 at Pattambi, Kerala

Njavara rice holds hope for cancer patients (2007, August 2007) *The Hindu* p.1

Padmaja, P., Geethakumari, V.L., Harikrishnan Nair, K., Chinnamma, N.P., Sasidharan N.K. and K.C.Rajan. 1994. A glimpse to problem soils of Kerala. Kerala Agricultural University, Thrissur.p.116.

Reddy G.S. 2000. Characterisation and Evaluation of rice (*Oryza sativa L.*) cultivar Njavara. Unpublished M.Sc. (Ag.) Thesis, Kerala Agricultural University, Thrissur, Kerala 137p.

Sahadevan, P.C. 1966. Rice in Kerala, Agriculture Information Service-Department of Agriculture, Kerala.

Sasidharan, N.K., Rajan, K.C. and Balachandran, P.V. 2003. *Rice ecosystems in Kerala. (eds.) Narayana Kutty, M.C., Prema, A. Jyothi, M.L., and Balachandran, P.V.*In National Symposium on " Priorities and Strategies for rice research in high rainfall tropic" organized by Regional Agricultural Research Station, Pattambi on 20-21, August 2004 at Pattambi, Kerala pp. 90-103.

Sasidharan, N.K. 2004. *Pokkali- The World acclaimed sustainable farming system model*. In : ICAR short course on Rice Fish Integration through organic farming for sustainablility and food security organized by Regional Agricultural Research Station, Pattambi on 1-10, December 2004 at Pattambi, Kerala p. 58-68.

Siddiq, E.A. 2004. *Aromatic and Medicinal Rices: Urgency for Collection, Validation and Conservation for Sustainable Utilization towards value added rice farming.* Keynote delivered at the Inaugural Session of the Science- Society Interface on 'Medicinal and Aromatic rices organized by Regional Agricultural Research Station, Pattambi on 20-21, August 2004 at Pattambi, Kerala.

Chapter 20

Rice Farming Transforms Farmer's Lives

C. Karthikeyan, K.A. Ponnusamy, N. Anandraja and T. Parthasarathi

Directorate of Extension Education,
Tamil Nadu Agricultural University
Coimbatore – 641 003, Tamil Nadu
e-mail: karthikeyanextn@yahoo.com

SUCCESS STORY – 1

System of Rice Intensification (SRI) technology breaks the jinx of conventional rice practice. Adoption of SRI has transformed his life, says **Thiru. P. Solaimalai** of Villiyankunram village, Madurai district who is a B.A. graduate. He started his life as office staff in mid day meal program at primary school, Andaman village, Madurai. He started farming on 13 acres of his land after getting inspired by various talks provided by department officials on Agricultural Science.

Thiru P. Solaimalai

The farmer switched over his primary occupation to agriculture. In the past 40 years he is practicing rice cultivation for two seasons in an year using Periyar Vaigai irrigation water. During 2001- 2006, he introduced SRI technology in his field for the first time in Madurai district. He followed SRI cultivation every year in his field and participated in state level crop harvest competition. He bagged the first prize in competition recording the highest yield of 24112 kg/ha receiving a cash price of Rs. 25,000/- from Honorable Chief Minister of Tamil Nadu during the year 2004-05.

During 2011-12, Chief Minister of Tamil Nadu made announcement to bring out second green revolution using SRI practice. The Madurai East circle Agriculture Officers Thiru N. Ramasamy, Thiru T.R. Sridhar and Assistant Agriculture Officer Thiru J. Subburaj demonstrated the SRI technology in Thiru P. Solaimalai's field as initiative. At the time of demonstration Assistant Director of Agriculture Thiru S. Kangaraj emphasized that cultivating CR 1009 rice variety in Samba season by SRI technology might give double the yield. He encouraged the farmer (Thiru P. Solaimalai) to attend the crop yield competition announced by Chief Minister. The positive inspiration from the ADA and previous price winning experience motivated the farmer to participate the contest.

The farmer followed the technology strictly and adopted proper cultivation practices as per the suggestions given by Assistant Director of Agriculture Thiru S. Kanagaraj. The following key practices were adopted to achieve higher yield.

Selection of Variety

The recommended CR 1009 variety seeds were procured from Pattukottai, Tanjavur District. The foundation seeds were used for sowing purpose.

Seed Treatment

The paddy seeds (3 kg/ac) were treated with *Pseudomonas* @ 10g/kg of seed. The next day again treated with 2 kg of *Azospirillum* and 2 kg of Phospobacteria treated seeds sown in raised bed nursery.

Main Field Preparation

During summer season, he practiced goat penning in the main field. Already planted daincha plants were incorporated in to the field before flowering.

He carried out the soil tests and followed recommended dose of fertilizers. He applied basal dose of fertilizers in the form of Urea (50 kg), DAP (100 kg), Potash (25 kg) and Zinc Sulphate (10 kg) and leveled the field properly. Including this, 20 packets each of *Azospirillum* and Phosphobacteria were mixed with 50 kg of farm yard manure and broadcasted the mixture uniformly in the main field.

Rice Planting and Management Practices in the Main Field

The 14 days old rice seedlings were transplanted using marker machines with a spacing of 22.5 x 22.5 cm following single seedling/hill as per the recommendation. Blue green algae @ 4 kg/acre mixed with sand and broadcasted in the field at 10 DAT.

Marks are made by Marking Machine

Weeding was done by conoweeder at 10, 20, 30 and 40 DAT. The conoweeder kills the old root on both the directions and helps to produce the newer roots, which leads to increase the number of tillers. The Urea 25 kg along with 5 kg neem cake and 25 kg potash were mixed and applied uniformly at 25 DAT. Second and third time, Urea and Potash were mixed at a rate of 25 kg each and applied evenly. Based on the crop growth, Urea was applied as per the Leaf Colour Chart (LCC) code.

Plant Protection

For controlling of leaf folder, 400 ml Nuvagran and 100 g Bavistin were mixed with 200 litre of water and sprayed at evening hours using hand sprayer.

Water Management

The farmer practiced alternate wetting and drying method of irrigation in his field as per the training provided on "water management" by Joint Director of Agriculture, Madurai.

Yield

The farmer had practiced conventional method of paddy cultivation previously and obtained yield of 8000 kg per hectare using ADT 39 variety. Adopting SRI technology, he obtain an increased grain yield due to its more grain weight (8,272 kg per acre *i.e.*, 20,680 kg per hectare).

Chief Ministers Special State Level Crop Production Competition First Price Winning Details through SRI Technology

1.	Crop	Paddy
2.	Variety	CR - 1009
3.	Area	2 Acre
4.	Sowing date	25.09.2011
5.	Seedling age	14 Days
6.	Planting date	08.10.2011
7.	Harvesting date	21.02.2012
8.	Age	149 days
9.	Area Harvested for competition	0.50 Acre
10.	Threshed grain weight (Wet)	4136 Kg
11.	Acre Yield (127 Bags; Each bag 65 Kg)	8272 Kg
12.	Hectare Yield	20680 Kg

Cultivation Cost Details (Hectare)

Cost Details		Rs.
Goat Shed (15 days; Rs. 200 per day)	:	3000.00
Farm Yard Manure 75 Carts (Rs. 150/cart)	:	11250.00
Daincha seeds 20 kg (Rs. 65/kg)	:	1300.00
Tractor Plough (2 Times)	:	4000.00
Sowing labour charge	:	200.00
Raised bed nursery preparation	:	800.00
Seed 8.0 kg (Rs. 30/kg)	:	240.00
Seed Treatment (*Pseudomonas, Azospirillum, Phospobacteria*)	:	40.00

Cost Details		Rs.
Main Field Preparation	:	
Tractor Plough (4 Times)	:	6000.00
Bund formation	:	1500.00
Levelling	:	750.00
Marker Rolling	:	800.00
Basal Fertilizer (2 Bag Urea, 4 Bag DAP, 1 Bag Potash)	:	4750.00
Paddy Micro nutrient fertilizer	:	925.00
Paddy transplanting	:	5000.00
Top dressing fertilizer (5 Bag Urea, 1 Bag Potash)	:	1650.00
Conoweeder weeding (10, 20, 30 and 40 DAT)	:	2400.00
Crop Protection Spraying (Labour charges Included)	:	4000.00
Irrigation	:	2500.00
Harvesting	:	5250.00
Cleaning and other charges	:	1145.00
Total	:	**57500.00**

The farmer attributed his success to decreased cost of irrigation, labour and cultivation besides increasing the cropped area by 50 per cent due to intermittent irrigation. It is of no surprise that he had been awarded with Prashasthi Patra and a cash award of Rs.1 Lakh by his Excellency, The President of India.

Further, he has received cash prize of Rs.5 Lakhs and medal by Hon'ble Chief Minister during the Independence Day function for his dutiful adoption of SRI.

Thiru P. Solaimalai was honored by Tamil Nadu Agricultural University as "Velan Sathanayalar Award" during 2012.

Contact Address

Thiru P. Solaimalai, B.A.

Kanjanampettai,

East Andaman

Velliyankundram

Madurai (Dt)

Mobile: 9344131977

SUCCESS STORY – 2

The System of Rice Intensification posses the primary procedure for paddy cultivation and has changed the livelihood of farmers. Mrs. T. Amalarani belonging to Vasudevan Nallur village of Thirunelveli District, says that the System of Rice Intensification has changed her life style. She is a B.A. Graduate and her interest towards farming have increased due to the efforts put in by the Agricultural Officers, she cultivated in her 10 Acre of land.

Mrs. T. Amalarani

She chose Agriculture as her primary profession. For the past 20 years during both the seasons she cultivate paddy through well and canal as the main source of irrigation. In, 2003 she had introduced the technology of System of Rice Intensification on her farmland and it was the first time for to be introduced in Thirunelveli district. She cultivated paddy by this method.

In the year 2011 - 2012 Honorable Chief Minister of Tamil Nadu has announced the high yielding competition with the objective of System of Rice Intensification. Mr. Ramasamy, Assistant Agriculture Officer in Sivagiri Taluk, Thirunelveli district has broadly explained about the System of Rice Intensification technology. Then he intimated the chance to obtaining double the yield by adopting this System of Rice Intensification technology by using Trichy-1 Rice variety during later Samba season. Moreover, he encouraged Miss. T. Amalarani to participate in the crop production competition as announced by the Honorable Chief Minister. As suggested by the

Assistant Director of Agriculture, she implemented the cultivation technology as per the recommendation.

Variety Selection

She procured the recommended Trichy - 1 variety from Rajapalayam. Foundation seeds were used as the seeding material.

Seed Treatment

Three kilo gram of paddy seeds per achre was treated with 10 g/kg of Psuedomonas and the day after these seeds were again treated with bio-fertilizers *viz.*, two kilo gram Azhosphirillum and 2 kilo gram of Phosphobacteria and sowed in a raised bed nursery.

Preparation of Land

At first, 50 tonnes farmyard manure was spread on the land. Then he ploughed the land with already grown garlic crop.

Based on the soil tests the following were applied as basal fertilizer. 50 kg of urea, 100 kg of DAP, 25 kg of Potassium and 10 kg of Zinc sulphate were applied as basal fertilizer and the land was leveled. In addition to this, 12 pockets of *Azospirillum* and 50 kg of Phoshobacteria was mixed equally with farmyard manure and broadcasted.

Maintenance of Paddy Cultivable Land

As recommended 14 days aged seedlings were transplanted (Spacing: 22.5 x 22.5 cm) in main field by using paid labors.

On the 10th, 20th, 30th and 40 days after sowing, conoweeder was used for weed management. The usage of Conoweeder discards older roots and promote the emergence of newer roots. Moreover weeds are used as fertilizer. 25 kg of Urea, mixed with 5 kg of Neem oil cake then 25 kg of Potash are added equally and broadcasted 20 days after sowing.

Crop Protection

Leaf roll caterpillar can be controlled by using 400 ml Novakran and 100 g Bevistin, it is mixed with 200 litres of water and sprayed using hand operated sprayer at evening hours.

Water Management

Water management practices were followed based on the training conducted by the Assistant Director of Agriculture.

Rice Field

Yield

Miss. T. Amalarani obtained an yield of 7000 kg of grains by following normal method of cultivation. After followed the System of Rice Intensification, the grain yield and additional grain weight were increased. As per above reasons she got the yield of 7257 kg grains per Acre which means 18143 kg grain yield/Hectare.

Rice Field

Details of Special State Level competition on 'System of Rice Intensification based crop production' announced by Honorable Chief Minister of Tamil Nadu

1.	Crop	Paddy
2.	Variety	Trichy-1
3.	Sowed Area	4 Acre
4.	Seedling Age	14 days
5.	Age between sowing and harvest	140 days
6.	Harvested area for competition	2.5 Hectare
7.	Yield per Acre	7257 Kg
8.	Yield per Hectare	18143 Kg

Cost of Production Details (per Acre)

	Cost Details	Rs.
1.	Farmyard manure 10 tractors (Tractor: 500)	Rs.5000
2.	Flatted garlic 10 kg (Rs. 35/kg)	Rs. 350
3.	Plouging tractor (Two times)	Rs. 1500
4.	Sowing cost (1 Man cost)	Rs. 300
5.	Preparation of raised bed nursery (2 Man cost)	Rs. 600
6.	Cost of 3 kg of paddy seed Trichy - 1 (Rs. 32)	Rs. 96
7.	Seed treatment (Pseudomonos-2 pack, Azospirillum-2 pack)	Rs. 40
8.	Sowing land preparation (Tractor plouging 4 times)	Rs. 1900
9.	Cost of canal preparation 3 Mans (Rs. 300/Man)	Rs. 900
10.	Levelling operation	Rs. 300
11.	Basal fertilizer	
	10 Kg Zinc sulphate + sand mixing for sowing	Rs. 1000
	1 Pack of DAP (Rs. 980) + ½ pack of Potash (Rs. 155)	Rs. 900
	Pseudomonas-10 packs + *Azospirillum*-10 packs with compost manure	Rs. 300
12.	Cost of sowing (sowing with holding of rope)	Rs. 1950
13.	5 Men for using conoweeder on 10, 20, 30 and 40 days (4 x 5 = 20 Mans and cost is 130/Man)	Rs. 2600
14.	Again basal fertilizer ½ packs of Urea (Rs. 130) + 25 kg of Neem oil cake (Rs. 400)	Rs. 530
15.	Sprays (2 times)	Rs. 1500
16.	Irrigation water	Rs. 1000
17.	Harvest cost (Machine/hour)	Rs. 1200
18.	Cost of labour (during harvest time)	Rs. 900
		Rs. 22986

She achieved the success through alternate irrigation, minimum labour, cost of cultivation and reduction of 50 per cent cultivable area. Miss. T. Amalarani received "Krishikarman" award and Rs. 1 lakh cash reward from the Honorable President of India.

Tamil Nadu Agricultural University has awarded and honored Miss. T. Amalarani as 'Agricultural Successor Award 2012' for her success.

For Contact

Miss. T. Amalarani,

Vasudeva Nallur,

Mainroad,

Sivagiri – Taluk,

Thirunelveli District.

Mobile: 9443969766

SUCCESS STORY - 3

Being an agriculturist, this was the moment when my heart felt with high satisfaction and happiness when I spoke to Mr. S. Sethumathavan, who is a successful farmer in SRI technology. He hails from A. Pudupatti, Alanganallur village, Madurai district practicing farming for the past 15 years.

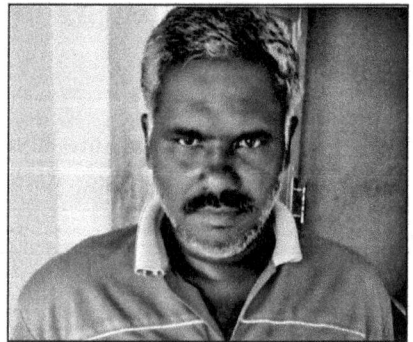

Mr. S. Sethumathavan

On speaking with him, he happily laughs and says "I was practicing ADT variety previously in 2002 with help of irrigation from Periyar Vaigai Irrigation Scheme. That year I achieved an average yield of 5500 kg/ha, but was not happy with the yield. I was thinking of adopting some technology that would provide me higher yield. That time I went for a meeting that was conducted by Agriculture Department. Mr. R.M.Lakshmanan (Alanganallur ADA) along with Mr.T.R.Sridhar (AO) and Mr.P.Alagu murugan (AAO) participated the meeting. They explained about the

SRI system and showed me a record-breaking achievement by a farmer, who won the prize from central and state government with a yield of 8000 kg by adopting SRI. I got excited and approached ADA about the technology and he instructed the officers to guide me. I adopted SRI technology with their help and cultivated CR - 1009 rice variety using 8 Kg of seeds".

Seed Treatment

5 Kg of seed was treated with 50 g of *Psuedomonas* (PF1) and dried for two days followed by treating with 2 packets of *Azospirillum* and *Phosphobacteria*. The treated seeds were sown in nursery.

Main Field

For main field preparation, farmer reared goat and used its dropping as manure. He used daincha as green manure for the field. Finally the main field was puddled for twice and during the last puddling, 30 trucks of farmyard manure was applied. Based on soil testing results, he applied the following chemical fertilizers:

1. DAP   –   125 kg
2. Urea   –   125 kg
3. Potash   –   80 kg
4. Zinc sulphate –   25 kg

After leveling the field, he applied 12.5 kg of Paddy Micronutrient, 20 packets of *Azozpirillum* and 20 packets *Phosphobacteria* mixed with FYM.14 days aged CR 1009 rice seedlings were transplanted at 25x25 cm spacing.

Weeding

The weeding operation was done by conoweeder at 10, 20 and 30 days after planting (DAP). Conoweeding provides nutrients if mulched in the soil.

Top Dressing Fertilizer

Based on the leaf colour chart, he applied nitrogen in form of urea mixed with two parts of potash as per recommendation.

Crop Protection

Imidacarpoloid 750 ml/ha mixed with 1 liter of humic acid was sprayed to control leaf roller/webber.

Water Management

Alternate wetting and drying irrigation method was carried out as suggested by Agriculture officer. The water level was maintained at 2.5 cm level at 10 DAP and 3 cm after weeding. The AOs projected yield of 2.5 kg/sq.mt - motivated the farmer to participate in state level competition. Under the judge's supervision, the crop harvested in the field yielded 24,002 kg/ha and the farmer was awarded the first prize at state level.

Details of Chief Ministers Special State Level Crop Production Competition First Price Winning information through SRI Technology

1.	Crop	Paddy
2.	Variety	CR-1009
3.	Area	2 Acre 45 Cent
4.	Sowing date	01.09.13
5.	Seedling age	14 Days
6.	Planting date	15.09.13
7.	Harvesting date	24.02.14
8.	Age	160 Days
9.	Area Harvested for competition	1 Hector
10.	Acre Yield	9600 Kg
11.	Hectare Yield	24000 Kg

Cultivation Cost Details (Hectare)

	Cost Details	*Rs.*
1.	Goat Shed (20 days; Rs. 200 per day)	4000.00
2.	Farm Yard Manure	500.00
3.	Daincha seeds 25 kg (Rs. 30/kg)	750.00
4.	Tractor Plough (1 Times)	1500.00
5.	Sowing labour charge	500.00
6.	Seed 16.0 kg (Rs. 40/kg)	640.00
7.	Tractor Plough (2 Times)	3000.00
8.	Bund formation	2500.00
9.	Levelling	1000.00
10.	Marker Rolling	5000.00
11.	Basal Fertilizer (1½ Bag Urea, 2 1 Bag DAP, 1½ Bag Sulphate)	4500.00
12.	Paddy transplanting	1050.00
13.	Conoweeder weeding (20,10)	3000.00
14.	Crop Protection Spraying (Labour charges Included)	600.00
15.	Irrigation	2250.00
16.	Harvesting	3000.00
17.	Cleaning and other charges	2400.00
	Total	**36190.00**

Finally the farmer concluded by saying " I adopted SRI technology in 2 acres 45 cents and produced 24 tonnes in 5 ½ months. SRI technology is one of the treasure provided by Agriculture department and I would recommend all the farmers to use it for higher production". They can use the following techniques to achieve a greater yield.

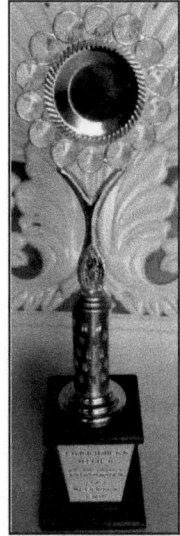

1. Fourteen day young seedlings should be used to produce more side tillers.

2. Integrated nutrient management involving both organic (Daincha, FYM and biofertilizers) and inorganic fertilizers reduces the overall cost of chemical fertilizer.

3. Practicing mechanized conoweeder for removing weeds will provide additional root growth and side tillers.

4. Planting space of 25x25 cm gives lesser competition within crop canopy for water, fertilizer, air and aids for reduced pest disease incidence and rat damage. This saves 100 bags of rice and cost of Rs.2000.

Contact Details

Mr. S. Sethumathavan

S/o. Chandran

A.Pudhupatti Post,

Alanganallur

Madurai District

Mobile: 9543844005

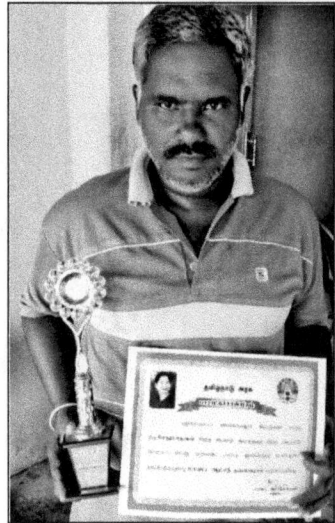

Index

Plate 2.2: Up Land Rice and Wetland Rice. (p. 42)

Plate 4.3: Domestication of Rice. (p. 43)

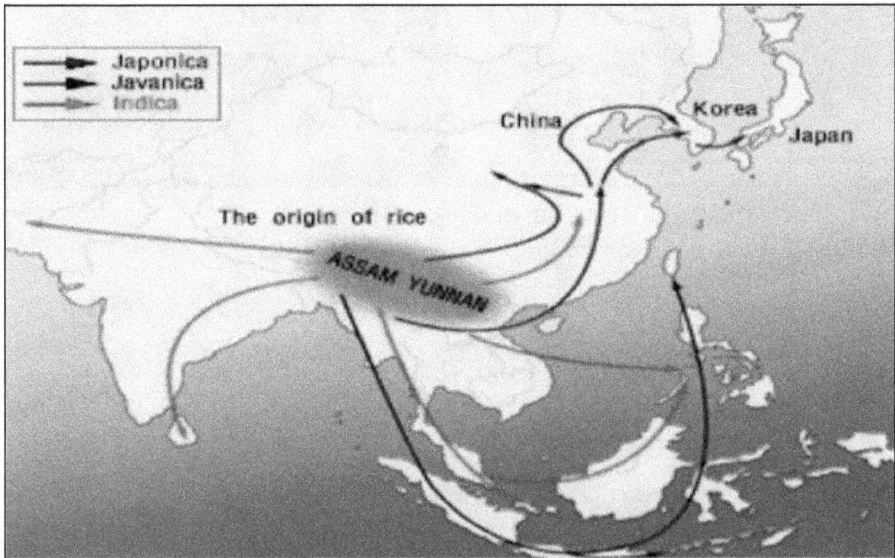

Plate 4.4: Early Spread of Rice. (p. 44)

Plate 4.5: Wild Rice. (p. 45)

Plate 4.7: Semi Dwarf Rice Varieties. (p. 48)

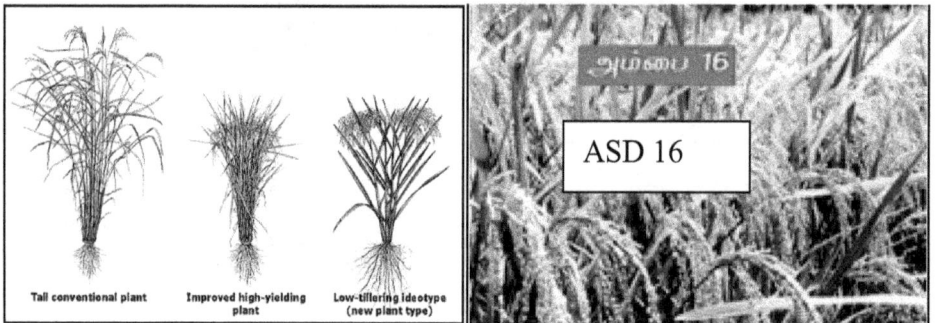

Plate 4.8: New Plant Type. (p. 49)

Plate 4.9: Super Hybrid Rice. (p. 50)

(p. 56)

(p. 59)

Wet-bed Nursery–Seeds Sowing (p. 63)   **Wet-bed Nursery–Pulling Out the Seedlings (p. 63)**

Dry-bed Nursery–Layout (p. 64)   **Dry-bed Nursery (p. 64)**

Dapog Nursery (p. 64)

Dapog Nursery–Seedling Mat (p. 64)

Modified Mat Nursery (p. 66)

Modified Mat Nursery–Seedling Mat (p. 66)

Tray Nursery (p. 67)

Machine Transplanting (p. 67)

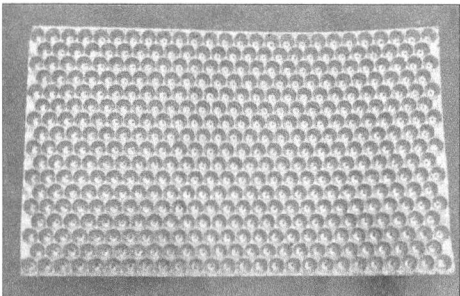

Bubble Tray (p. 68)

Bubble Tray Nursery–Seedlings (p. 68)

Figure 7.1: In the Omission Plot where N has not been Applied, Leaves are Yellowish Green. (p. 75)

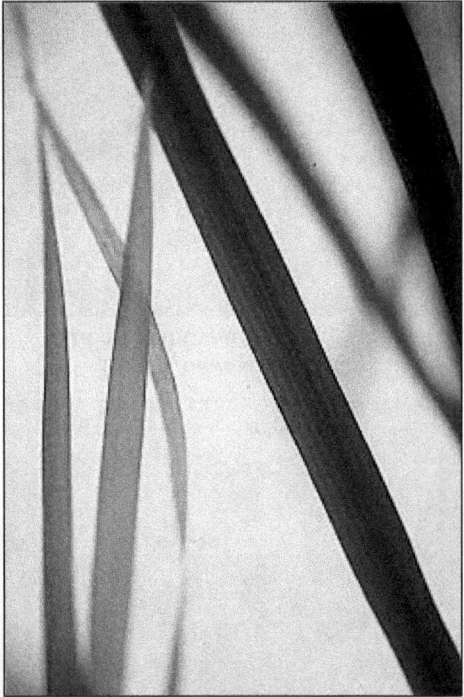

Figure 7.3: Tillering is Reduced where P is Deficient. (p. 77)

Figure 7.2: In N-deficient Plants, Leaves are Smaller. (p. 76)

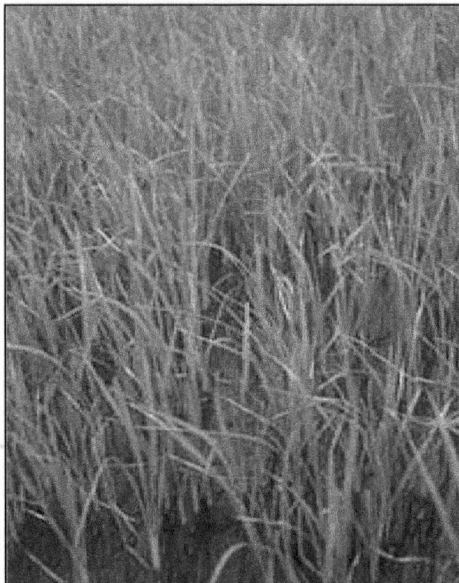

Figure 7.4: Even Under Less Pronounced P Deficiency, Stems are Thin and Spindly, and Plant Development is Retarded. (p. 77)

Figure 7.5: Plants are Stunted, Small, and Erect Compared with Normal Plants. (p. 77)

Figure 7.6: Leaf Margins become Yellowish Brown. (p. 78)

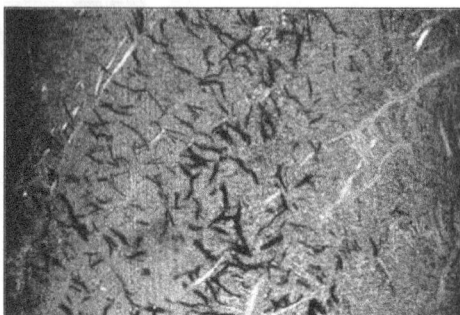

Figure 7.7: K-deficient Rice Plant Roots may be Covered with Black Iron Sulfide. (p. 78)

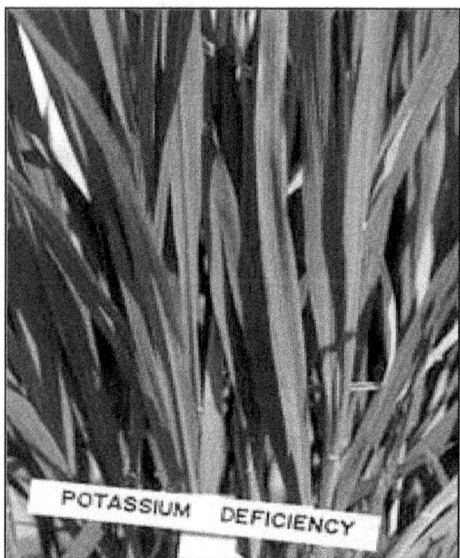

Figure 7.8: Leaf Bronzing is also a Characteristic of K Deficiency. (p. 79)

Figure 7.9: The Leaf Canopy Appears Pale Yellow Due to Yellowing of the Youngest Leaves and Plant Height and Tillering are Reduced. (p. 80)

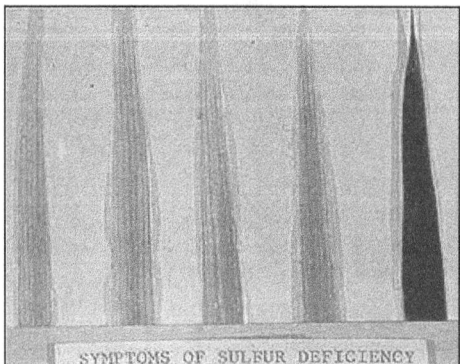

Figure 7.10: Chlorosis is more Pronounced in Young Leaves, where the Leaf Tips may Become Necrotic. (p. 80)

Figures 7.11: Symptoms Only Occur Under Severe Ca Deficiency when the Tips of the Youngest Leaves may Become Chlorotic-White. (p. 81)

Figure 7.12: Uneven Field Growth, Plant Stunting (Foreground). (p. 82)

Figure 7.14: Deficiency is mainly a Problem on Upland Soils. (p. 84)

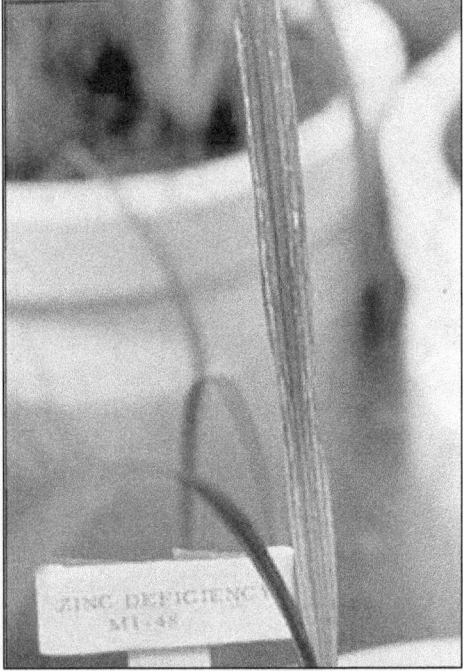

Figure 7.13: Appearance of Dusty Brown Spots on Upper Leaves. (p. 83)

Figure 7.16: Deficiency mainly Occurs in Organic Soils. (p. 85)

Figure 7.15: Symptoms Appear as Interveinal Yellowing of Emerging Leaves. (p. 84)

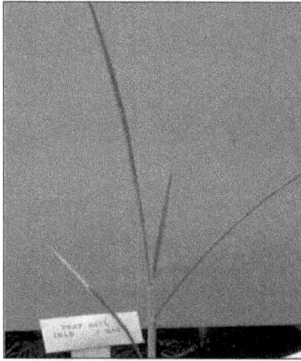

Figure 7.17: New Leaves may have a Needlelike Appearance (p. 86)

Figure 7.18: Brownish Leaf Tips are a Typical Characteristic of B Toxicity, Appearing as Marginal Chlorosis on the Tips of Older Leaves. (p. 87)

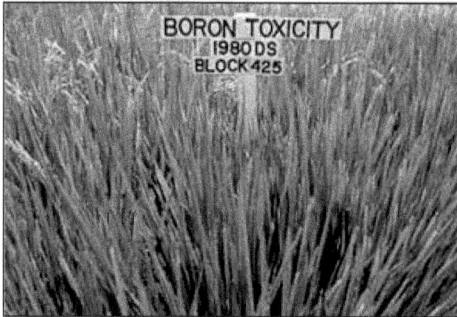

Figure 7.19: 2-4 Weeks Later, Brown Elliptical Spots Develop on the Discolored Areas. (p. 87)

Figure 7.20: Rice Growth is Characteristically Patchy in Soils Affected by Salinity. (p. 88)

Figure 7.21: Where Saline Irrigation Water is Used, Patches of Affected Plants are found Adjacent to Water Inlets. (p. 88)

Figure 7.22: Plants are Stunted with White Leaf Tips. (p. 88)

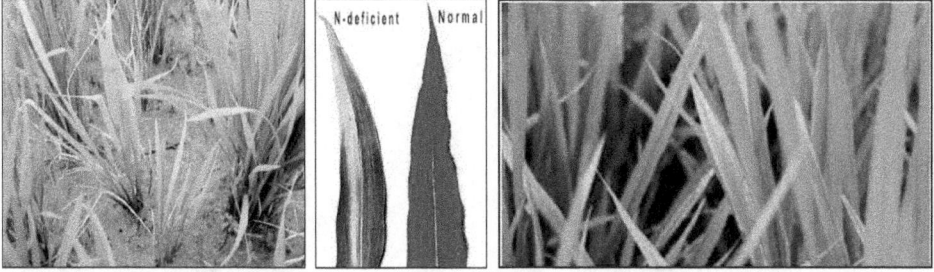

Plate 12.1: Symptoms of Nitrogen Deficiency. (p. 147)

Plate 12.2: Symptoms of Phosphorus Deficiency. (p. 148)

Plate 12.3: Symptoms of Potassium Deficiency. (p. 149)

Plate 12.4: Symptoms of Sulfur Deficiency. (p. 150)

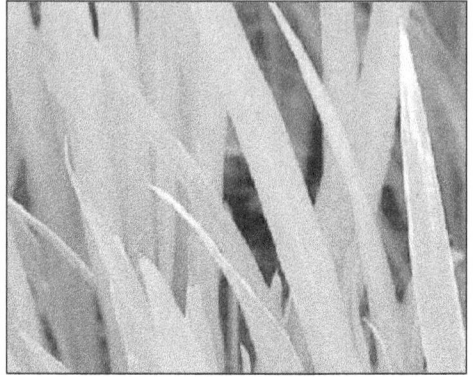

Plate 12.5: Symptoms of Calcium Deficiency. (p. 151)

Plate 12.6: Symptoms of Calcium Deficiency. (p. 152)

Plate 12.7: Symptoms of Iron Deficiency. (p. 153)

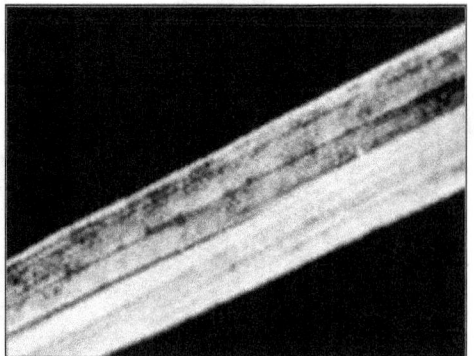

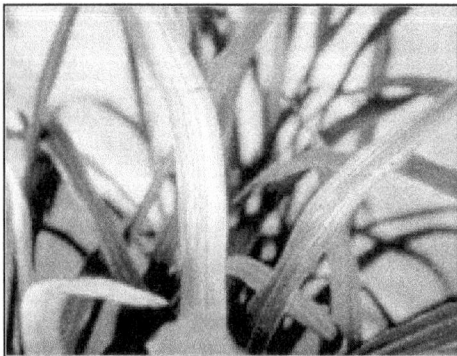

Plate 12.8: Symptoms of Manganese Deficiency. (p. 154)

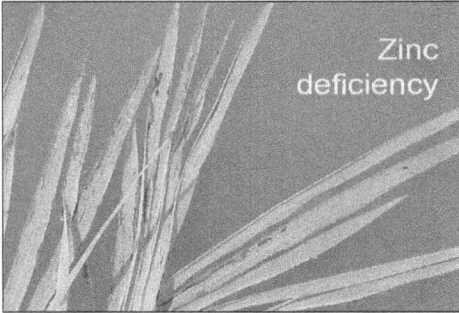

Plate 12.9: Symptoms of Zinc Deficiency
(p. 155)

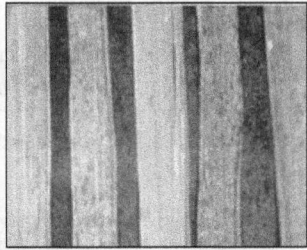

Plate 12.10: Symptoms of
Boron Deficiency. (p. 156)

Plate 12.11: Symptoms of
Copper Deficiency. (p. 157)

Plate 12.12: Symptoms of
Silicon Deficiency. (p. 158)

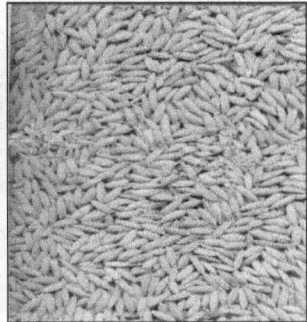

Figure 14.1: Rice Inflorescence. (p. 182)

1.5 kg of salt dissolved in 10 lts. of water.
Partial floating of egg
(size of 25 paise coin above solution)

Separation of seed

Figure 14.3: Egg Floatation Technique for Paddy Seed Upgradation. (p. 186)

Figure 14.6: Grinding of Sprouts with Cold Water. (p. 188)

Figure 14.5: Cow Pea Sprouting. (p. 188)

Figure 14.7: Squeezing. (p. 189)

Figure 14.8: Nursery Sowing. (p. 190)

Figure 14.9: Nursery Weeding. (p. 190)

Figure 14.10: Field Preparation. (p. 192)

Figure 14.11: Transplanting. (p. 192)

Figure 14.12: Transplanting Sequence for 8:2 (A x R) Row Ratio. (p. 193)

ROW RATIO – F : M = 8 : 2 **(p. 194)**

Figure 14.14: Rope Pulling. (p. 197)

Figure 14.15: Rod Driving. (p. 197)

Figure 14.16: Rogueing. (p. 198)

Figure 14.17: Removal of Off-type. (p. 198)

Figure 14.18: Abnormality of Panicle. (p. 199)

Figure 14.19: Purple Tip. (p. 199)

Figure 14.20: Split Husk. (p. 200)

Figure 14.21: Wild Rice – Objectionable Weed. (p. 200)

Figure 14.22: Damage by Rice Hispa. (p. 202)

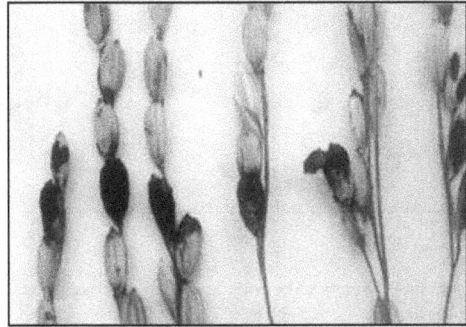

Figure 14.23: Symptom of Paddy Leaf Blast. (p. 203)

Figure 14.24: Symptom of Paddy Bunt. (p. 203)

Figure 14.25: Harvesting Male Row (First Removal). (p. 204)

Figure 14.26: Harvesting Female Row. (p. 204)

(p. 208)

(p. 210)

(p. 211)

(p. 212)   (p. 213)

(p. 213)

(p. 216)

(p. 216)

(p. 217)

(p. 218)

(p. 219)

(p. 220)

(p. 221)

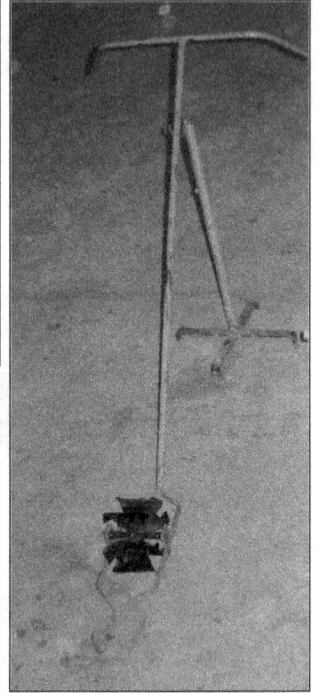

(p. 221-223)

(p. 225)

(p. 226)

(p. 226)

(p. 227)

Figure 17.1 (p. 231)

Figure 17.2 (p. 233)

Figure 17.3 (p. 234)

Figure 17.4 (p. 234)

Figure 17.5 (p. 236)

Figure 17.6 (p. 237)

Figure 17.7 (p. 238)

(p. 240)

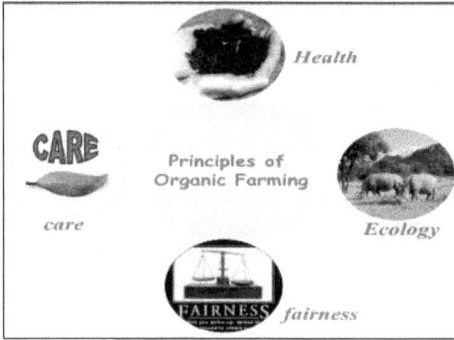

(p. 240)

(p. 243)

(p. 244)

(p. 244)

(p. 245)

(p. 245)

(p. 245)

(p. 246)

(p. 247)

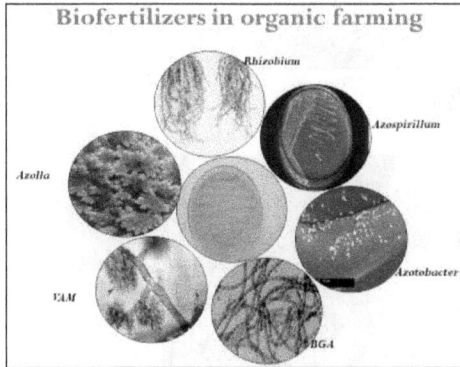

(p. 248)

(p. 248)

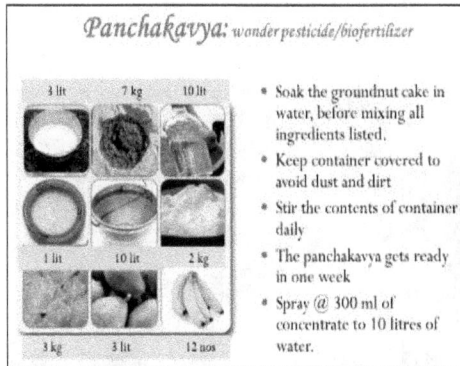

(p. 249)

(p. 250)

(p. 250)   (p. 251)

(p. 252)

(p. 253)

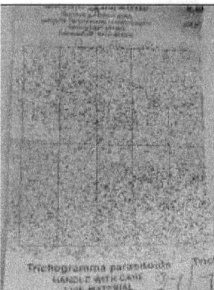

(p. 254)

Palakkadan Matta- Jyothi (p. 261)

Pokkali rice- Vytilla 1 (p. 261)

Kaipad Rice (p. 263)

Jeerakasala Rice (p. 264)

Jeerakasala Rice (p. 264)

Gandhakashala Rice (p. 264)

(p. 270)

(p. 271)

(p. 271)

(p. 272)

(p. 275)

Rice Field

Rice Field

(p. 276)